Calf Husbandry, Health and Welfare

John Webster
Professor of Animal Husbandry
University of Bristol
School of Veterinary Science

GRANADA
London Toronto Sydney New York

Granada Technical Books
Granada Publishing Ltd
8 Grafton Street, London W1X 3LA

First published in Great Britain by
Granada Publishing 1984

Published in the United States of America in 1984 by
Westview Press, Inc., 5500 Central Avenue, Boulder, Colorado 80301

ISBN (USA only): 0-86531-760-7

British Library Cataloguing in Publication Data
Webster, John
Calf husbandry, health and welfare.
1. Calves
I. Title
636.2'07 SF 205

ISBN 0-246-11910-1

Typeset in Great Britain by Cambrian Typesetters, Aldershot, Hants
Printed and bound in Great Britain

Contents

Acknowledgements vi
Preface ix

1 **Normal development** 1
Problems at birth – the development of digestion, temperature regulation and resistance to disease – standards for welfare.

2 **Nutrition and digestion** 17
The nutritive value of feeds, energy, protein, minerals, vitamins – feed composition – digestion in the rumen and abomasum – nutrient requirements.

3 **Feeding systems from birth to weaning** 46
Bucket-feeding teat-feeding – composition of milk replacers, starter rations – performance and health of calves on different systems.

4 **Environmental needs** 71
Thermal comfort and temperature regulation – physical comfort and space requirements, environment and health.

5 **Housing design, ventilation and disinfection** 98
Simple shelters – monopitch calf houses – ventilation of calf houses – controlled environments – cleaning and disinfection.

6 **Common diseases and their recognition** 118
Signs of health and disease – recognition of disease causing generalised ill health – diseases of the digestive and respiratory tracts – skin and other localised diseases.

7 **The development of behaviour** 144
Oral activity, rest and play behaviour, 'vices', response to man – effects of different rearing sytems on behaviour.

8 **Production sytems and economics** 165
Calf production from suckler cows – spring v. summer-calving – husbandry systems for veal calves – contract calf-rearing – the economics of disease.

9 **Stockmanship and calf welfare** 187
Stockmanship – markets and transport – starting the 'bought-in' calf – routine operations – 'the reasonable bounds of humanity'.

Index 201

Acknowledgements

I gratefully acknowledge my enormous debt to farmers, stockmen, scientists and veterinary surgeons too numerous to mention who have contributed to my education. In particular I thank my graduate students who have done more than anyone, not only to keep me up to date, but also to fire my enthusiasm for new ideas and improved understanding. I thank Dr. C.D. Mitchell and Claire Saville for permission to reproduce (and adapt) illustrations, and my especial thanks go to Pauline Webber who typed and retyped the manuscript.

Preface

This book is intended primarily for the good stockman. To him or her I say at the outset that it contains a lot of straight science but only, I believe, as much as is necessary to provide a logical basis for the practical advice that follows. I hope too that it will be useful to undergraduate and graduate students of agriculture and veterinary science. Although I have tried to reflect current scientific understanding as accurately as possible, I cannot, in a book of this size, include long, comprehensive references to scientific articles, so most of the suggestions for further reading relate to books and reviews. Many of the topics under discussion have, as yet, no clear scientific explanation. In these cases I use phrases like 'I think' or 'in my experience' to convey a sense of caution rather than dogma, i.e. to indicate the way things seem to be in the light of present knowledge.

The book deals, in sequence, with the basic principles of husbandry, feeding, housing, health, behaviour, economics and welfare. Each section begins with an explanation of relevant aspects of calf function, e.g. digestive physiology, temperature regulation, development of behaviour and then evaluates conventional and novel husbandry systems in terms of their effects on animal performance, health and welfare. These basic principles apply, of course, wherever in the world calves are reared. Only in dealing with matters such as legislation have I had to be parochial and restrict discussion to law in the United Kingdom or European Economic Community.

Throughout this book I have attempted to analyse methods of calf-rearing mainly from the point of view of the calf and ultimately it is they who will, I hope, derive most benefit from it.

1 Normal development

Husbandry is a good word. When applied to any aspect of farming it easily incorporates all facets of scientific and economic production but it also overlays these things with two essential senses, the senses of care and preservation. The sense of care implies care for the individual animal or crop. The sense of preservation applies to the flock, the herd and ultimately to preservation of the land. There is, furthermore, nothing wishy-washy or 'back to nature' about good husbandry. The land and the stock are there to be worked, and worked hard, and the harder they are worked the better the husbandry has to be. I am reminded of one of the oldest and best of all agricultural jokes. The Minister of the Kirk is passing the immaculate allotment of one of his parishioners: runner beans ten feet high, Brussels sprouts regular and erect as soldiers, not a weed in sight.

'That's a fine display John, that you and the Good Lord have created.'
'Aye Minister, but ye should ha' seen it when the Guid Lord had it tae hissel.'

This book is concerned with the rearing of calves to 6 months of age, after which, in natural circumstances, they might expect to be self-sufficient. The natural circumstances for a calf are, of course, to be born on range, or at pasture, and to run with its mother for this time. On the whole, the beef cow on open range makes a pretty good job of rearing her own offspring, provided, of course, that due care is paid to her husbandry and she is kept well nourished in a healthy environment. Indeed, if all calves were reared by their mothers there would be little need for this book.

The calf born to the dairy cow, however, is routinely submitted to more insults to normal development than any other farm animal. It is taken from its mother in the first four days of life. It is deprived of its natural first food, whole cow's milk, and fed one of a variety of cheaper liquid substitutes for milk. However, since milk replacers are still very expensive compared with solid foods in which the energy comes principally from carbohydrate, the calf is usually weaned onto these solid foods as quickly as possible.

It is also normal practice to transport dairy calves from their farms of origin onto specialist rearing units. Such calves often travel long distances and may, during the course of their travels, pass through one or more markets. This practice not only submits the calf to the rigours of travel but denies it the opportunity to feed normally and exposes it to a high risk of infection. In view of all these insults it is obvious that the first essential of good husbandry in rearing the calf born to the dairy cow is to keep it alive and fit enough to perform well later on. Other aspects of good economic husbandry, like minimising costs of feeding, housing and labour, will of course assume equal importance in time, but for the moment they can wait.

Mortality and disease

Table 1 summarises the results of several surveys of calf mortality in the period immediately after birth and then during the first 3–4 months of life. Here and elsewhere in this book, no attempt has been made to present an exhaustive survey of published information. Those seeking a more bibliographic introduction to scientific research into calf health and husbandry would do well to consult books like *The Calf* (Roy, 1980) or *Le Veau* (Mornet and Espinasse, 1977).

The observations presented in Table 1 are sufficient to illustrate several vital points. Most surveys published before 1980 put calf mortality in the U.K. at about 5%; it was higher for bought-in calves than for home-reared calves and was at its highest in large specialist units rearing calves for beef or veal. This is no reason to infer that husbandry is worse in large specialist units, merely that it is more difficult. Recorded death rates of calves born to dairy cows in the large dairying areas of the U.S.A., California and Michigan, reach levels of 20% or greater. However, a very recent study conducted by the U.K. State Veterinary Service in association with the University of Bristol revealed that in 72 selected calf-rearing units mortality was, on average, only about 1%. These figures do not include deaths occurring within the first 2–3 days of life before the calf is moved to the rearing unit. Stillbirths and neonatal deaths would probably have accounted for about another 2% (Allen and Kilkenny, 1980).

Table 1 Mortality rates in calves from birth to about 6 months of age; summaries of various surveys.

	Details	Mortality (%) 1 d–6 months	0–1 day
United Kingdom			
Withers (1952, 1953)	England	6	
	Scotland	12	
Leach and others (1968)	Home-reared	5.0	5.4
	Purchased calves	8.0	
Milk Marketing Board (1964)	Dairy calves,		
	home-reared	6.3	
	purchased	13.5	
Kilkenny & Rutter (1975)	Suckler beef units	3.6	1.9
	Purchased calves,		
	specialist units	5.8	
	mixed farms	4.6	
State Veterinary Service (1983)	Suckler beef units	0.5	
	Home-reared	0.6	
	Purchased calves	1.3	
United States of America			
Oxender and others (1973)	Dairy herds, Michigan,		
	< 100 calves	10.8	6.2
	> 100 calves	14.3	7.7
Martin and others (1975)	Dairy herds, California, each approx. 300 cows	20	

Footnote:
Kilkenny & Rutter (1975), in *'Perinatal Ill-health in Calves'* Ed. J. M. Rutter, Compton, Berks.; Leach and others (1968), *MAFF Animal Diseases Surveys* Report No. 5; Martin and others (1975), American J. Vet. Research **36**: p. 1099; Milk Marketing Board (1964), Report of the Breeding and production organisation, **14**: p. 100; Oxender and others (1973) J. Amer. Vet. Med. Assoc. **162**: p. 458; State Veterinary Service (1983), unpublished results.

It appears, however, that calf mortality in the U.K. is much less than it used to be, although even when it was high it was nothing like as high as that which has been reported from the big dairy areas of the U.S.A. The reason for this is quite simple. When the value of calves is high, death rates are low. The marked increase in the value of calves born to dairy cows in the U.K. has come about largely in response to an increased efficiency of production of beef from the dairy herd and this has largely followed the march of the British Friesian to a position of total dominance in the national dairy herd. Speaking from a purely aesthetic point of view, the demise of the Ayrshire and Channel Island breeds is a shame, because they are pretty cows and there is no reason to assume that they are any less efficient at converting grass to milk than the British Friesian. In the context of calf welfare, however, the demise of the pure dairy cow must, on balance, be a good thing, since it has also brought about the virtual demise of the 'bobby calf,' the male calf from the dairy herd which, having no value, tended to receive precious little care and attention and spent a nasty, brutish and short life moving through dealers' yards to early slaughter. The large number of 'bobby calves' coming especially from Ayrshire cows goes a long way to explain the high calf mortality rates observed in Scotland in the late 1940s (Table 1). It equally explains the high death rates of dairy calves in the U.S.A. They had no value as beef or veal and so they were allowed to die.

The rearing of calves for quality veal has received a lot of criticism, both rational and irrational, on the grounds that it imposes unacceptable insults to the welfare of the animals. I discuss this matter in detail in Chapter 8. At this stage, however, I would point out that veal production is essentially a system for exploiting calves which are surplus to requirement for both milk and beef production. The essential difference between the calf bought for quality veal and that bought for bobby veal is that, with the former, the buyer has a vested interest in keeping it alive and healthy and good health has to be the single most important criterion of sound welfare.

A dairy calf's chances of surviving the first few weeks of life depend therefore on its potential value as a dairy replacer, or, if it is a male, on its potential for beef (or possibly veal). Before leaving this point it may be worth reminding the British dairy farmer that although the British Friesian came to prominence almost entirely because of the amount of milk she produces, she has also, to date, been an excellent dual-purpose cow producing British beef, nearly 60% of which comes from the dairy herd. The North American Holstein looks a very attractive creature when considered simply in terms of milk production, but she evolved in an agricultural environment where the great majority of beef originates from the cow on open range. Faced by this competition, the male Holstein calf has little value and so its beefing characteristics have assumed little importance. Now I do not want to go any deeper into the everlasting argument about the relative merits of British Friesians and Canadian Holsteins. In fact, well before it is settled, there will probably no longer be any difference. All I am saying here is that when it is sensible to produce beef from the dairy herd it is also sensible to include beefing characteristics in dairy cow selection.

Purchased calves are, on average, five times more likely to require treatment for disease, especially respiratory disease or 'scours', than calves reared on their farm of origin. Clearly the husbandry of bought-in calves is rendered a good deal more difficult by the stresses of transport and exposure to infection that they suffer in early life. It is probably also fair to say that all bought-in calves suffer to some degree from disease in the first weeks of life. The 50% or more which require antibiotic and other forms of treatment are those in which disease is present in such a severe form that without treatment they would either die or at least experience a chronic set-back to normal growth. It is a good working rule to assume that all calves bought in from dealers are incubating one or more infections and should be nursed accordingly.

The increase that has occurred in recent years in the value of calves, especially male calves born to cows from the dairy herd, has promoted an enormous upsurge in interest in the development of 'better' methods for rearing them. In this context, 'better' can describe scientific improvements in calf-rearing based on improved understanding of the physiology, health and behaviour of the animals. It can also simply mean cheaper. For example, a marked reduction in feed costs resulting from the use of 'milkless' milk replacers could save the farmer money even if such a system increased mortality rate slightly. I must point out right away that this is a theoretical example. There is no evidence to show that properly formulated 'milkless' milk replacers increase mortality (see Chapter 3). However, long gone are the days when calf-rearing was simply a matter of putting a calf into a clean, individual pen and feeding it twice daily from a bucket. A wide range of alternative calf-rearing systems is now offered to the farmers and, human nature being what it is, the advocates of each lay claims that their particular system is the best. The advocate of individual pens and buckets says that 'a calf's worst enemy is another calf', and that hygiene is the first essential of good husbandry and, up to a point, he is right. The advocate of rearing calves in groups with free access to milk through a nipple feeder says that his system is less likely to cause digestive upsets to the young calf because it is much closer to the natural state; the calf drinks what it wants when it wants, and he too is right – up to a point.

In this book I shall deliberately refrain from recommending 'best ways' to rear calves. All sorts of combinations of feeding, housing and management can be successful in the right hands and on the right farm. Moreover, a system that works well on one farm may fail on another for reasons that can be inexplicable even to the expert. The approach that I shall take is to examine, in a scientific way, the most important aspects of calf physiology, health and behaviour during the first few months of life and explore the impact on these things of different husbandry practices. This should enable the calf-rearer to develop a system of husbandry which successfully reconciles the needs of the animals with the particular circumstances of his or her own farm, or, if things go wrong, to modify the husbandry in a fundamentally sound way in order to put things right.

Normal development – the problem of birth

It is necessary, therefore, to begin by examining the normal development of a calf born to a beef cow on pasture on the basis of the adage 'a calf's best friend is its mother'. This is, in fact, not always such a truism as it sounds. The traditional beef cow may be excellently designed for rearing calves but the modern, highly selected Friesian dairy cow is a most unsatisfactory mother on several counts. She eats too much for a suckler cow, she gives far too much milk for one calf and her udder is altogether the wrong shape, so that the newborn calf often has extreme difficulty in finding a teat since they tend to be much closer to the ground than he has been programmed to expect. The crossbred beef cow, containing 50% or more Friesian blood, is also open to the same criticisms, albeit to a lesser degree.

However, let us assume a calf has been born to the right sort of cow. What I shall do now is describe in very simple terms the sequence of events as they occur during normal development. This description, albeit superficial, inevitably presupposes some knowledge of the anatomy and physiology of the calf. Each of the organ systems will be considered in greater detail in subsequent chapters and these will, I hope, explain the omissions, both simple and complex, that are inevitable in this general introduction.

A calf's first problem is to get itself successfully born. The subject of obstetrics is outside the scope of this book but there are some very important aspects of husbandry which affect the calf's chances of surviving the delivery process. Table 2 shows how the incidence of stillbirths and assisted calvings from beef suckler cows is affected by sire breed (Allen and Kilkenny, 1980). This quantifies something that most good stockmen are already aware of in general terms, namely, that the bigger the bull relative to the size of the cow, the greater the probability of problems at calving. The table does, however, say rather more than just that. Both the Hereford and Aberdeen Angus are 'safe sires'. The advantages of the Aberdeen Angus over the Hereford are trivial, which means that both breeds are good sires for small cows and heifers. It is, of course, up to the individual breed societies and organisations such as the Milk Marketing Board to ensure that individual bulls within these or any other breeds are rejected if they, as individuals, appear to contribute to a significant increase in calf mortality. The high incidence of mortality and assisted calvings for the offspring of Charolais, Simmenthal and South Devon bulls may not be entirely due to their large size but to the shape of their head and shoulders as well. It appears, furthermore, that when Charolais or Simmenthal sires are used on British breeds of cow, pregnancy must be slightly prolonged. This inevitably results in a bigger calf and increases the risk at calving.

Table 2 Sire breed effects on calving problems in suckler cows (Allen and Kilkenny, 1980)

Sire breed	Assisted calvings %	Surgical calvings %	Calf mortality %
Angus	2.4	0.1	1.6
Charolais	9.0	1.2	4.6
Devon	6.4	0.3	3.6
Hereford	4.0	0.3	2.0
Limousin	7.4	0.7	3.8
Lincoln Red	6.7	0.3	3.2
Simmenthal	8.9	1.2	4.4
South Devon	8.7	0.9	4.2
Sussex	4.5	0.2	2.1
Welsh Black	3.8	0.2	2.6

The high incidence of calving problems in cows mated to large bulls, especially the Charolais, is a worry in itself yet it is only the tip of the iceberg. If there are basic genetic reasons for presupposing that calving may be difficult, it also follows that for every one cow that delivers a stillborn calf as a result of obstetric complications, perhaps another 6 or 7 will have abnormally prolonged deliveries which can lead to anoxia in the calf and some degree of brain damage. In time this may be completely reversible but there are a lot of things the calf has to do very early on for which it needs a clear head.

The importance of colostrum

It is absolutely essential for the newborn calf to drink an adequate amount of the cow's first milk – colostrum or 'beestings' – ideally as soon as possible after birth and certainly within the first 18 hours of life. This colostrum carries the maternal antibodies essential to protect the calf from the infections it will experience in early life, before it has generated its own active system of immunity. There is copious evidence to show that calves which fail to drink sufficient colostrum early on have a greatly increased risk of succumbing to infection (Brignole and Scott, 1980). It is essential for them to get this colostrum as soon as possible after birth because it is only at this time that the gut wall is sufficiently permeable to absorb the maternal antibodies unchanged. The permeability of the gut to these antibodies starts to decline very shortly after birth and has almost disappeared by about 18 hours. The calf that is weak and stupid as a result of a long and difficult delivery is unlikely to drink sufficient colostrum during this time and this imposes a severe threat to its subsequent health and welfare.

To return for a moment to the matter of sire selection. A number of good stockmen have told me that (especially) Charolais calves are variously 'dreamy, stupid, poor doers or disease-prone.' This is, I think, a fair reflection of what the stockmen have observed but it is a little unfair to what is, in the right circumstances, an excellent beef sire. What these stockmen have been observing are the immediate and longer term consequences of anoxia during calving. The good stockman has to be aware of the probability of calving problems, particularly in the case of suckler cows at pasture, and must ensure, wherever possible, that calves are able to suck freely from their mothers in the first hours of life. If this is not practicable, either because the calf is too weak or too stupid to suck, or because the design of the cow (typically the old Friesian) is such that the calf *cannot* suck, then it is essential to take other steps to ensure that it has adequate colostrum. If the calf is subsequently going to have to drink from a bucket, then it might as well start now. The suckler's calf or the weak calf should be drenched 3–4 times in the first 18 hours with 1 litre (about 2 pints) of colostrum, ideally through a tube placed in the oesophagus (see Chapter 6) to avoid any material getting into the lungs and setting up an inhalation pneumonia.

Digestion

Colostrum is, of course, not just antibodies, it is food. The first instinct of a newborn calf is to rise to its feet and suck from its mother to obtain, first from colostrum and later from milk, the nutrients essential for its early growth and development. Later on, it will begin to eat grass and other plant materials containing cellulose which can only be utilised by the animal after undergoing prior microbial fermentation. The digestive system of the young calf (or any young ruminant) is beautifully designed to ensure efficient digestion of a high quality feedstuff like milk and simultaneously the efficient fermentation of plant materials containing cellulose and other structural carbohydrates. Figure 1 illustrates in very schematic form the anatomy of the stomachs and small intestine of the calf shortly after birth.

On anatomical grounds the ruminant stomach is divided into four parts; reticulum, rumen, omasum and abomasum. For practical purposes it is easier to consider it as two functional units: the reticulo-rumen, which is concerned with microbial fermentation of plant materials, and the abomasum, which is comparable to the true stomach of humans or pigs (for example), and is concerned with acid digestion (Fig. 1). The omasum remains something of a mystery but probably not a very important mystery. Food passes down the oesophagus which opens into the anterior dorsal sac of the rumen. Solid food eaten by a ruminant of any age is then propelled over the anterior pillar into the main body of the rumen by contractions of the reticulum and anterior dorsal sac. Contractions of the rumen mix the food with salivary and other juices in the presence of ruminal micro-organisms and fermentation takes place. As the fermentation process proceeds, liquids and the smaller, lighter particles which are

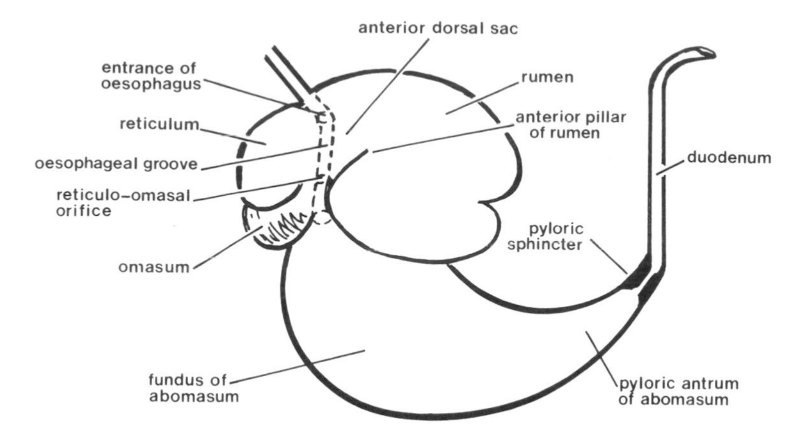

Fig. 1. A schematic illustration of the forestomachs, stomach and duodenum of the newborn calf.

made up mostly of rumen microbes and highly fermented material rise to the top, are carried back over the anterior pillar and pass on into the abomasum via the reticulo-omasal orifice. As indicated already, this fermentation process is essential for the utilisation of plant cell walls. It is, however, wasteful, or even dangerous, to submit milk or other highly nutritious materials to rumen fermentation. Fortunately the ruminant has a beautiful mechanism, the oesophageal groove, which, during normal early development, allows milk to pass direct into the abomasum where it can be subjected to the highly efficient process of enzymic digestion. The oesophageal groove is, as its name implies, a groove or channel in the wall of the anterior dorsal sac. Directly above this channel is a specialised area of muscle. When the calf sucks milk from its mother, a nervous reflex is stimulated which causes this specialised muscle to contract in such a way that it entirely closes the 'roof' of the oesophageal groove, converting it into a pipe which delivers milk straight from the oesophageal opening through the reticulo-omasal orifice, down the short, clear channel at the base of the omasum and into the abomasum or true stomach. The calf which runs at pasture with its mother for the first six months of its life is able therefore to direct all food to where it will be best utilised: solid food to the reticulo-rumen and milk to the abomasum.

When milk enters the abomasum it initially forms a clot under the action of the enzyme rennin. (Calf rennin or 'rennet' is used to make junket or 'curds and whey'.) Most of the protein and nearly all the fat is contained within this clot. The whey passes fairly rapidly out through the pyloric sphincter for further digestion and absorption in the small intestine. The clot or curd is digested slowly in the abomasum and the contents are released equally slowly into the duodenum.

During the first week of its life the calf may drink 7–10 times a day from its mother and it will seldom consume more than 1 litre at a session. In normal circumstances therefore the calf is unlikely to overload the abomasum at any one time, and the digestive process in the abomasum is such that the flow of nutrients to the duodenum and small intestine is regulated at a steady rate. All this makes for good digestion and reduces the risk of the calf succumbing to enteritis, leading to diarrhoea (or 'scours'), or enterotoxaemia. These two conditions may be defined thus: enteritis is disease of the small intestine, enterotoxaemia is disease originating in the small intestine but also liberating toxins or poisons which invade the entire body.

Very shortly after birth, calves will begin to investigate with their mouths grass and all sorts of other objects in their environment. Early on, they spend a lot of time just playing with grass etc., but they soon get the idea of what is and is not edible and by two weeks of age they can be eating considerable amounts of solids.

It is, of course, also necessary for them to take in from the grass and the soil the micro-organisms which will colonise the developing rumen and ferment the grass. The steadily increasing intake of food, and particularly the nature of the fermentation process itself, stimulate the rumen to develop rapidly in size and function so that by about three months of age the volume of the reticulo-rumen has increased to about 25–30 ℓ and accounts for about 80% of the total volume of the complex ruminant stomach. Meanwhile, of course, milk is still passing directly into the abomasum through the oesophageal groove. This dual food-processing system is essential to normal development in a calf at pasture since, even though the rumen is fully functional at about 3 months of age, the quality of nutrients provided by the grass is inadequate to sustain normal growth at this time.

The whole question of the development of rumen digestion and nutrition will be dealt with in much more detail in Chapter 2. This section has been included simply to introduce the subject of how and why a calf feeds as it does during normal development. It should be obvious already that this feeding behaviour and physiology is very different from that which is imposed by many rearing systems, e.g. bucket-feeding once a day to weaning at five weeks. The implications of artificial feeding systems for calf growth and calf health will be discussed in Chapter 3.

The next problem to consider is how the calf comes to terms with the fact that it has left the secure, controlled environment of the uterus and now has to contend with the outside world. Some of the most elegant questions of neonatal physiology are those which concern the onset of respiration and the cardiovascular changes associated with the shift in oxygen and carbon dioxide exchange to the lungs from the placenta. This subject has been covered very well by Comline (1973). I shall not deal with it here because it is something that nearly all calves manage very well for themselves without help and if they don't, there is little that the stockman or the veterinary surgeon can reasonably be expected to do about it.

Temperature regulation

The problem of temperature regulation in the first days of life, however, depends very much on the environment into which the calf is born and there is plenty that the stockman can do about that. This will be discussed in Chapter 4. At this stage it should be sufficient to say that the newborn calf has a very mature capacity to regulate its body temperature and combat the stresses of cold compared with the newborn puppy or piglet for example, both of which would, in the natural state, be born and raised in nests which provide warmth and shelter.

The cow's first task is to lick the calf dry. This is very important not only in establishing the bond between mother and offspring, but it also reduces the risk of the calf becoming severely chilled through excessive heat loss by evaporation of the foetal fluids. Again, if the cow is prevented from licking her calf dry for any reason, the stockman should be prepared to give the calf a good rub down, especially in cold weather. The newly-delivered calf is acutely vulnerable to cold stress in the same way as a child coming out of a swimming pool, and for the same reason. Once it is dry and has had its first feed, its ability to resist cold becomes very impressive, again provided that its coat does not become saturated with moisture from rain or snow or excessively muddy conditions underfoot.

Out of doors then, a young calf will be stressed by cold from time to time although not a great deal in most circumstances. Problems of cold (and heat) stress can of course be reduced by the provision of fairly simple shelters.

Health

The great advantage of the outdoor environment for the calf is that it is clean, in the sense that the density of infectious micro-organisms is very low indeed compared with the situation that exists in most cattle houses. Chapter 6 deals with the common diseases of calves. The most common diseases undoubtedly fall under the two general headings of 'scours', and pneumonia. It is fair to say that neither of these disease complexes can be considered as straightforward, predictable responses to infection by pathogenic micro-organisms. Calves are frequently infected by organisms known to be associated with scours or pneumonia and yet will show no clinical symptoms of disease. Equally, calves may incubate a pathogenic organism for days or weeks without harm and then, apparently as the result of some environmental stress, suddenly fall sick. These important diseases are best considered as a complex interaction between the host, the parasite and the environment. Here, the host is the calf, the parasite may be one or more of a whole range of infectious micro-organisms and the environment includes anything that may affect the magnitude of the challenge from the parasite or the ability of the host to mount an adequate defence against that challenge (Webster, 1981).

Out of doors with its mother, the calf is subject to a number of environmental stresses but in most circumstances the magnitude of the challenge by infectious organisms is low and it is within the capacity of the calf's immune system to mount a defensive response. Here again, the calf running at foot with its mother on its farm of birth is at a decided advantage. Not only has it received antibodies in colostrum but those antibodies have been synthesised in the cow against the infectious agents that exist on that farm. As a general rule, one can say that the further a calf is taken from the environment in which its mother lives, the less resistance to disease it can hope to derive from its mother's colostrum.

A field full of cows and their calves cannot really be said to be all that hygienic. However, provided both cow and calf are adapted to that environment, the calf will receive a steady and tolerable exposure to infection. At first, this is under the specific protection of maternal antibodies, but later the calf develops its own active immune mechanisms to combat these common infections. That is normal development for a healthy calf. Nevertheless, even under these apparently ideal conditions some calves will fall sick or drop dead.

Behaviour

The behaviour of any animal is determined both by genetics and environment. At birth, its behaviour must be entirely determined by genetics. How it develops thereafter will be modified to a greater or lesser extent by the environment which it experiences. When we use the word 'normal' in the context of animal behaviour and behavioural deprivation, it becomes a much more loaded expression than when it is used to describe, say, digestion or metabolism. It is fair to say that the development of behaviour in a calf at pasture with its mother should be reasonably normal. If that is so, then one might reasonably say that the behaviour of a three-week-old calf bedded on straw in a comfortable, individual pen and bucket-fed twice a day was abnormal. Very few people would, however, describe such a husbandry system as unacceptable. It is an acceptable environment provided that it can be shown that the calves have successfully adapted to that environment and their behaviour, though different from that of calves at pasture, is normal to that environment and provided that there is no evidence to suggest they are suffering any unacceptable degree of distress. Chapter 7 will deal in some detail with the development of behaviour in different environments and attempt to resolve at least some of the thorny questions of what constitutes normal and abnormal behaviour.

Welfare and 'Animal Rights'

The Brambell Report on the Welfare of Farm Animals kept in intensive systems (1965) proposed, amongst other things, that all farm animals should be permitted five basic freedoms: 'freedom, without difficulty, to stand up, lie down, turn round, stretch their limbs and groom themselves'. These relatively modest proposals (which are not met in many intensive systems of livestock production) relate of course only to behavioural deprivation, which cannot be considered the sole insult to animal welfare. If, for example, a farmer decided to let his pre-weanling calves out of their individual pens to run and play together and if as a consequence half of them contracted a fatal case of enteritis, he could not honestly claim to have improved their welfare. The revised Codes of Welfare for Cattle, prepared by the Farm Animal Welfare Council (H.M.S.O. 1983) states in its preface that 'the first essential of a good husbandry system is that it should ensure the health and behavioural needs of the animals'. This, very properly, recognises that good health is the first essential of good welfare. The Farm Animal Welfare Council, in a statement of policy and philosophy, has produced a more complete definition of satisfactory welfare in terms which might be called The New Five Freedoms:

Freedom from (1) hunger and malnutrition
(2) thermal and physical discomfort
(3) injury and disease
(4) suppression of 'normal' behaviour
(5) fear and stress

These are, I propose, a very powerful set of standards by which to evaluate any husbandry system, intensive or extensive. As an illustration, let us consider a system of husbandry which is not uncommon, and which has not, to my knowledge, come under serious fire from the protagonists of animal rights. Consider a flock of pregnant Clun Forest ewes overwintered in the U.K. on turnips planted in poorly drained land, with no access to shelter and close to a housing estate. Such animals would almost certainly be malnourished with respect both to energy and protein. They would frequently be wet, muddy and cold. The risk of foot injuries and infections would be very high and they would be at constant risk of, at least, fear of and at worst, death from marauding dogs. Set against this, their freedom to 'stand up, lie down, turn around, stretch their limbs and groom themselves' must be deemed small compensation.

During the course of this book I shall be considering a wide variety of permutations of systems of feeding, housing and management of calves in the context of the physiology, economic production, health and behaviour of the animals. It is up to the reader to reach his or her own decision as to whether any particular system meets the 'New Five Freedoms'. To close this introductory chapter but to begin the book proper, I would say that no system can be deemed perfect. The best possible system is, however, that which best reconciles the objectives of economic production, good health and sound welfare. The achievement of these aims, or even the desire to achieve these aims, is the essential quality of good stockmanship and that is what this book is all about.

Further reading

Allen, D. & Kilkenny, B. (1980) *Planned Beef Production.* Granada Publishing, London.

Brambell, F.W.R. (1965) 'Report of the technical committee to enquire into the welfare of animals kept under intensive livestock husbandry systems.' CMND.2836. H.M. Stationery Office, London.

Brignole, T.J. & Scott, G.H. (1980) 'Effect of suckling followed by bottle-feeding colostrum on immunoglobulin absorption and calf survival'. *Journal of Dairy Science*, **63**: pp. 451–456.

Comline, R.S. (1973) *Foetal and Neonatal Physiology.* University Press, Cambridge.

Farm Animal Welfare Council (1983) *Code of recommendations for the welfare of cattle.* H.M.S.O., London.

Mornet, P. & Espinasse, J. (1977) *Le Veau*, Maloine, S.A. (ed. Paris).

Roy. J.H.B. (1980) *The Calf*, 4th ed. Butterworths, London.

Webster, A.J.F. (1981) 'Weather and infectious disease in cattle.' *Veterinary Record*, **108**: pp. 183–187.

2 Nutrition and digestion

The first essential of good husbandry is proper nutrition. The science of nutrition may be considered under three headings:

(1) *The nutritive value of foods.* This is determined by the chemical composition of the food, measured in terms of the essential nutrients: carbohydrates, proteins, fats, minerals and vitamins.
(2) *The digestive capacity of the animal.* This determines the ability of the animal to render nutrients present in the food available for the processes of metabolism.
(3) *The nutrient requirements of the animal.* These depend on the size and the physiological state of the animal. In the case of the calf, nutrients made available by digestion are required for the maintenance of life and for synthetic processes associated with growth.

The proper feeding of calves therefore involves selecting feeds of the correct nutritive value and feeding them in the correct quantities in order to meet the nutrient requirements for growth as precisely and as cheaply as possible. Any discussion of practical feeding systems must therefore begin with an outline of the principles of nutrition.

The nutritive value of feeds

The basic analysis of any food into its principal nutrients and their destiny in the body of a young ruminant animal are illustrated in Fig. 2. The first thing to know about any foodstuff is its dry matter (DM) content, since it is this that contains the nutrients. (The water requirements of calves will be considered separately.) Milk typically contains only about 125 g DM/kg, hay about 800–850 g DM/kg and compound feeds 900 g DM/kg. The DM content of pasture grasses varies according to season and recent rainfall, the average being about 200 g/kg.

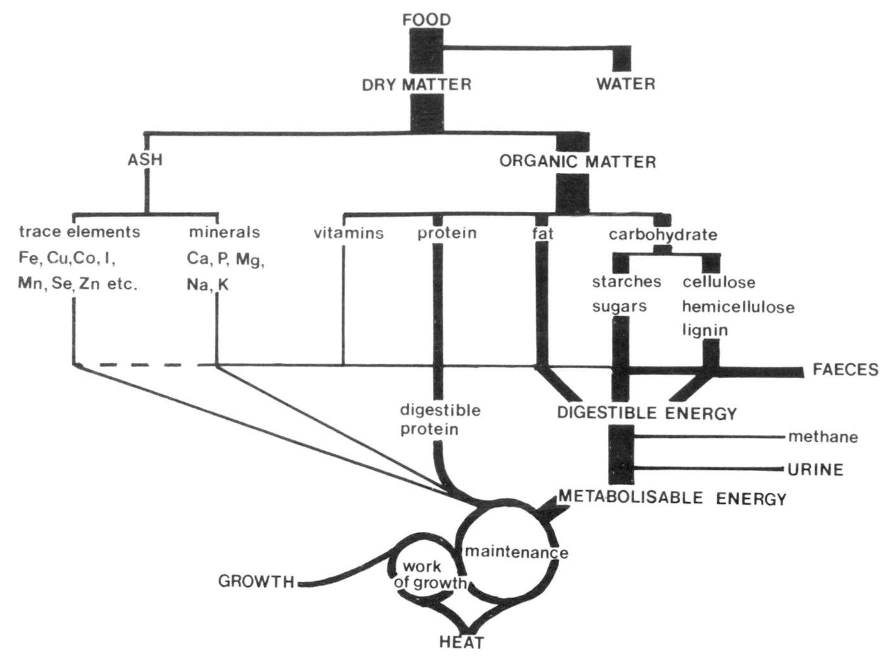

Fig. 2. The principal nutrients in food and their utilisation for maintenance and growth.

Dry matter is further subdivided into organic matter, carbon-containing, combustible materials, and ash, which consists mainly of the major inorganic elements like calcium, phosphorus and magnesium, plus very small amounts of trace elements like copper, cobalt, etc. (Fig. 2).

ENERGY

The conventional breakdown of organic matter is into carbohydrates, fats, proteins and vitamins. This is not, in fact, a particularly useful form of analysis in a nutritional sense, especially in the case of the ruminant, as it doesn't reveal much about the capacity of the food to meet the nutrient requirements of the animal. Overwhelmingly the greatest of the nutrient requirements is energy. In the young animal, the energy provided by the controlled combustion of organic matter through the pathways of intermediary metabolism does the work necessary to maintain life and support growth.

In a ruminant animal, for example the post-weaning calf, the principal source of food energy is carbohydrate. The carbohydrate in plants comes essentially in two forms: (a) starches and sugars, which are contained inside the plant cells; and (b) cell wall materials, the structural carbohydrates hemicellulose, cellulose and lignin. Plant cell walls cannot normally be broken down by acid digestion and if they remain unbroken it is not possible for the digestive juices to get at the starches and sugars inside the cell either. The ruminant animal first subjects plant cell walls to microbial fermentation in the rumen, which ruptures the cells, releasing the contents, and then digests most of the hemicellulose, some of the cellulose, although none of the lignin.

If foods rich in structural carbohydrates are to be fed to simple-stomached animals like man, pig, horse or the very young calf, the cell walls must first be broken down by physical processing ('rolling or grinding') or chemical treatment (e.g. sodium hydroxide) or by cooking. There is a simple laboratory analysis, the MAD fibre (modified acid detergent fibre) technique, which provides a reasonable estimate of the cellulose and lignin content in animal feedstuffs, based on their insolubility in boiling acid detergents. The MAD fibre content of a feed is a measure of its indigestibility for ruminant animals.

Fat makes a significant contribution to the energy value of the diet of a calf drinking milk but it normally plays little part in the nutrition of the post-weaning calf since it cannot be broken down in the anaerobic environment of the rumen. Moreover, it tends to inhibit the normal processes of microbial fermentation.

The Gross Energy (GE) value of a food is determined by measuring the amount of heat liberated when it is completely burnt at high pressure and in an atmosphere of oxygen in a bomb calorimeter. The same amount of energy (or enthalpy) is potentially available to the animal. However, losses of chemical energy occur at several stages of digestion and metabolism. Digestible energy, or more correctly 'apparently digestible energy,' is the difference between the energy content of the food and the faeces. It is called 'apparently digestible energy' since the faeces are not simply made up of undigested food but also contain microbial residues and endogenous material lost from the wall of the gut. In the ruminant there is a further loss of fermented energy from the rumen and large intestine, principally in the form of methane which is an essential end product of the anaerobic ('without oxygen') fermentation of carbohydrate. The amount of energy that passes out of the gut and into the blood and lymphatic system of the calf is therefore Gross Energy – (faeces

+ methane energy). A small amount of this absorbed chemical energy is lost in the urine; the remainder acts as the physiological fuel for the body and is called *Metabolisable Energy* (ME).

An animal which is neither gaining nor losing weight, nor producing milk, is said to be 'at maintenance'. When an animal is at maintenance with respect to energy, ME is entirely converted to heat as it supports the work of maintaining the essential organs of the body. ME fed in excess of maintenance can be used to support growth and fattening in the young animal or milk production in the adult female. This too requires work; ME in excess of maintenance cannot be retained in the body or body secretions with 100% efficiency. Readers who wish to pursue further the important intricacies of energy metabolism would do well to begin with that clear and elegant book *The energy metabolism of ruminants* by Blaxter (1967).

The reason why energy is the most important nutrient is illustrated in Table 3, which describes the fate of food energy in a variety of cattle offered a range of diets. The apparent digestibility of milk or milk replacer in the pre-weaning or veal calf is about 94%, i.e., energy loss in the faeces corresponds to 6% of intake. There is a further small energy loss in the urine but about 90% of GE appears as ME. In the young calf gaining about 1.3 kg/d on this highly nutritious diet, energy retention (RE) is only about 20% of GE; fully 70% is lost as heat. The veal calf growing very rapidly on a highly nutritious and highly expensive diet of milk replacer only manages to retain about 26% of GE. The conventionally reared calf at 6 months of age eating hay and concentrate food loses 47% of GE as faeces, urine and methane and only retains 11% in the growing body. The beef cow at maintenance retains no GE by definition; all is lost as chemical energy or heat. The high-yielding dairy cow retains about 26% of GE, a value comparable to that of the veal calf. However, she is, of course, a much more efficient creature in practical terms. Her overall efficiency of utilisation of ME (RE/ME) is 43% (26/61) as against 29% (26/90) for the veal calf. The cost of providing 1 MJ of ME from silage and concentrates is less than 1/5 of the cost of milk replacer and all the milk she produces is potentially available for human consumption as against perhaps 40% of the liveweight gains of a veal calf.

The reason why food energy is the most important and the most expensive constituent of any animal diet is that the efficiency of food energy conversion into animal product is so low. Practical

Table 3 The fate of food energy in cattle

	Diet	Weight gain (kg/d)	Gross energy (MJ/d)	Destiny of gross energy of food (%)				
				Faeces	Urine & methane	ME	Heat	Retained energy
Calf, 2 weeks old	Milk	0.3	14.1	6	4	90	70	20
Veal calf, 14 weeks old	Milk replacer	1.2	54.6	6	4	90	64	26
Calf, 6 months old	Hay & concentrate	0.6	69.8	35	12	53	42	11
Beef cow at maintenance	Hay only	0	58	45	10	45	45	0
Dairy cow, 30 ℓ milk/d	Silage & concentrate	0	352	25	14	61	35	26

ration formulation for farm animals is therefore based first and foremost on the principle of regulating the composition and the quantity of food offered so as to meet energy needs as precisely (and therefore as cheaply) as possible. Requirements for the other nutrients, amino acids, minerals and vitamins, can then be met by adjusting the concentration of these things in the ration so that no single nutrient is limiting. The reason that amino acids, minerals and vitamins are required in much smaller amounts than energy is, of course, that they can be recycled within the animal. Once ME has done work, it is lost as heat. The cost of providing dietary protein to meet the amino acid requirements of the animal comes a poor second to the cost of providing energy, but it is still expensive so it is important that protein nutrition is accurate too. Minerals and vitamins need to be present in such small proportions that one can, on the whole, afford to be generous.

PROTEIN

Dietary protein is broken down by digestion into its individual components, the amino acids. Once absorbed, these amino acids are resynthesised by the calf into proteins essential to maintenance and growth. The most obvious site for protein deposition is, of course, muscle, where the protein has a structural and functional role in posture and movement. However, muscle is a relatively inactive tissue, particularly in the food animals. Most protein synthesis takes place in tissues like the liver and the gut wall which are actively concerned with processing nutrients to meet the requirements of the body. These metabolic functions include such things as the synthesis of enzymes and hormones, cell division and cell repair, and so require a continuous supply of amino acids and energy. Most of the amino acids can be recycled but there is inevitably some loss of nitrogen each time a protein is degraded and this may be excreted via the faeces or the urine.

The conventional approach to the evaluation of animal feeds is to measure their N-content, multiply this by 6.25 on the assumption that N makes up 16% of most food proteins, and refer to this as Crude Protein (CP). For many feeds this can safely be assumed to be a measure of their capacity to provide amino acids, but in certain ruminant feeds such as silage or commercial feed supplements, especially liquid feed additives or feed blocks, much of the crude protein may be in the form of non-protein nitrogen (NPN), usually simple organic compounds such as urea. Adult ruminants on poor quality pasture or forage can benefit from NPN but, for reasons that will be developed below, NPN has no place in the feeding of calves under six months of age. If anyone is considering feed supplements for young calves it is essential to ensure that all the crude protein is true protein.

The nutritive value of food protein is too complex a subject to go into here. I would recommend for further reading *Animal Nutrition* (1981) by McDonald, Edwards and Greenhalgh. At this stage it is necessary to make only two points:

(1) Amino acids may be divided into the *essential* and *non-essential*. Essential amino acids (e.g. lysine, threonine, methionine, etc.) are those which must be provided in the diet because the animal is unable to synthesise them itself. Non-essential amino acids are those which the body can make for itself.
(2) Although the essential role of amino acids is protein synthesis, they can also act as energy sources, which means, of course, that they will be irreversibly oxidised to CO_2, H_2O and urea (Fig. 2).

The nutritive value of a particular protein source, e.g. soya bean meal or fish meal, depends therefore on the following:

(i) The quantity of true protein in the dry matter.
(ii) The composition of the protein in terms of its essential amino acids.
(iii) The absolute requirement of an animal for essential amino acids.

The first thing that determines absolute requirement for amino acids is the rate at which the body is producing new tissue in growth or lactation. Protein requirements for maintenance are small relative to energy requirements, because of amino acid recycling. As growth rate (or lactation rate) increases, so protein requirement increases both in absolute terms and in relative terms with respect to energy requirement. Thus a rapidly growing young calf will need a much higher protein content in its overall diet than, say, a yearling steer being stored over the winter at a very low growth rate.

The nearer the composition of a food protein approaches that of the protein in body gains, the more efficiently it can be used for growth since the supply of amino acids will be closely matched to requirement. There is little value in a food with a high protein content if this consists of an excess of non-essential amino acids, an inappropriate balance of essential amino acids or an absolute deficiency of a single amino acid such as lysine. There are several ways of evaluating the biological value of proteins for animals (see McDonald and others, 1981). Essentially, they compare the relative abilities of different proteins to promote growth in circumstances where amino acids are marginal or limiting. If a protein is fed in excess of requirement for the specific purpose of promoting body protein deposition, the amino acids will be used as sources of energy and lost. Protein will therefore be used inefficiently if the overall protein concentration of the diet is too high or if amino acid supply is inappropriate to the animal's needs. It should be obvious, therefore, that the main reason why feeds based on animal protein (fish meal or meat and bone meal) are so valuable and so expensive is that their amino acid composition is very close to that of the animal that eats them. However, the cheaper sources of vegetable protein (soya bean meal, etc.) may be only slightly imbalanced with respect to certain so-called 'limiting' amino acids and it is usually possible to make good a slight deficiency of, say, lysine by the addition of a small quantity of a lysine-rich protein source such as fish meal. There are other, non-nutritional reasons why one has to be careful when using sources of vegetable protein like soya bean meal in diets for calves, but this can wait until the next chapter.

MINERALS

It goes without saying that a proper diet for a young calf should contain sufficient minerals to promote normal growth. However, most practical diets designed to meet the calf's requirements for the principal nutrients, energy and protein, happen to contain a sufficiency of most minerals to meet these requirements too. Mineral nutrition only acquires practical importance if the principal constituents of the diet are deficient in a particular mineral or if some other element of the diet seriously reduces the availability (in effect, the apparent digestibility) of that mineral. Ruminant animals which must rely for most or all of their food on pasture or conserved forage are potentially subject to a wide range of mineral deficiencies (see Underwood, 1980). In most systems for artificially rearing calves to six months of age, minerals are usually provided in abundance and mineral deficiencies are rare.

The two mineral elements needed in greatest quantities by the growing calf are calcium and phosphorus. Most of the calcium and phosphorus is retained as calcium phosphate which is the main structural constituent of bone. Both elements also have other, more dynamic roles in animal physiology and metabolism. Calcium ions are involved in muscle function. The paralysis seen in cows which suffer milk fever shortly after calving is caused by an acute fall in calcium in the body fluids as calcium enters the milk faster than it can be absorbed from the gut or mobilised from bone.

Phosphorus is, amongst other things, an essential constituent of the compound adenosine triphosphate or ATP, which is the energy currency of the body. In simple terms, oxidation of absorbed dietary energy (ME) or body energy reserves (principally fat) is linked to a phosphorylation process which adds an energy-rich phosphate bond to ADP (adenosine diphosphate) connecting it to ATP. The chemical energy captured in this bond is the principal food for the work of the body (muscular exercise, nerve transmission, active secretion and absorption of chemicals etc.). Since phosphorus is so intimately linked to energy metabolism, it follows that the first signs of phosphorus deficiency are those associated with a general running down of energy metabolism, i.e. poor growth and poor appetite. These symptoms are not very specific but they are rather more helpful than some of the classic text book symptoms such as depraved appetite (gnawing bones) or even sudden death ('falling disease'). Animals would have to be rather far gone before such symptoms became apparent and the good stockman should have suspected that something was amiss long before then.

This illustrates a general point with regard to deficiencies of the minor nutrients (minerals and vitamins). Very often the earliest signs of deficiency are not very specific – the animals simply don't appear to be doing well. The stockman who waits for classic symptoms to appear, e.g. depigmented areas or 'spectacles' round the eyes in copper deficiency, has probably waited too long. It is usually possible for the stockman and his veterinary surgeon to make a tentative diagnosis of a mineral deficiency after eliminating other possibilities such as parasitism or infectious disease, if they know exactly what has been fed and have a reasonable, approximate knowledge of what minerals might be lacking, or out of balance in that diet. It is also, as a general rule, possible to diagnose a mineral or vitamin deficiency as it is treated since, if the animal is provided with the element it lacks, it will get better. If it doesn't, then something else is wrong. This sounds like a blinding glimpse of the obvious but it is rather more than that. An animal which acquires an infectious disease may reasonably be expected to get better without treatment since its natural defence mechanisms are designed to eliminate the source of infection. If one gives the infected animal an antibiotic, one cannot be sure whether the antibiotic has contributed to the cure or not. If an animal is faced by an absolute deficiency of a particular nutrient it *can't* get better unless that supply of that nutrient is increased.

The calcium and phosphorus contents of some calf feeds are listed in Table 4. As indicated already, the possibility of a Ca or P deficiency in artificially reared calves is remote. The only calves which could be at risk from these (and several other) deficiencies are the calves of suckler cows. This may occur either on poor winter pasture, which is deficient in most minerals but especially P and magnesium, or if the calves are kept in byres with their mothers in the way once popular in areas like the north-east of Scotland, and fed on milk, hay or straw and roots. Mineral deficiencies that can occur in these circumstances include phosphorous, magnesium, selenium, copper and iron and several vitamin deficiencies are also possible, but the hazards of this system are now fairly generally recognised and most farmers deem it expedient to supplement such a feeding system with some or all of these minerals, ideally in a controlled and palatable form, by incorporating them into a pelleted feed.

This raises another very important general point. Contrary to popular belief, cattle have no 'mineral wisdom'; that is to say, they have no idea what are their mineral requirements, nor how they should meet them. There is a single exception to this rule. Cattle can sense a sodium deficiency and can moreover smell the presence of sodium in rock salt, or in drinking water at quite incredibly low concentrations. However, sodium deficiency is seldom a practical problem for the healthy calf although it acquires crucial importance in scours. Otherwise, cattle simply have no more innate ability than we have to tell what mineral is what. When calves were experimentally made deficient in a variety of minerals and then offered a range of minerals free-choice, their intakes bore no relation to their needs. The practical conclusion to be drawn from this is that while it is very convenient to provide minerals for cattle free-choice in the form of blocks or powders, one cannot expect the animals to deal with them sensibly. Again, it has been shown that when minerals are offered free-choice to cattle in the field, perhaps 75% may be consumed by only 25% of the animals, 50% will eat a little and 25%, who may be the most in need, will not come near them. When-

Table 4 Approximate nutritive value of some feeds for calves

	Colostrum	Milk	Barley	Maize (corn)	Soya bean meal	Pasture grass	Ryegrass silage	Meadow hay	Straw (winter barley)
Dry matter (g/kg)	220	125	860	860	900	200	250	850	860
Composition of DM (g/kg)									
NFE (digestible carbohydrate)	140	368	795	825	360	485	390	495	390
MAD fibre	nil	nil	75	35	85	240	350	400	590
Crude protein	681	288	110	100	505	175	172	89	37
Lipid	163	280	17	42	16	40	40	16	16
Apparent digestibility (%)	93	93	86	87	73	65	61	57	39
Gross energy (MJ/kg DM)	20.6	23.6	18.2	18.9	19.6	18.0	18.4	17.6	17.8
Metabolisable energy									
(MJ/kg fresh material)	4.0	2.6	11.8	12.2	11.1	2.0	2.3	7.1	5.0
(MJ/kg dry matter)	18.0	20.6	13.7	14.2	12.3	10.0	9.3	8.4	5.8
Minerals									
Calcium (g/kg DM)	11.8	10.4	0.5	0.3	2.5	5.5	4.5	4.5	4.4
Phosphorus (g/kg DM)	10.9	8.8	3.7	3.1	11	2.8	2.8	2.1	1.0
Magnesium (g/kg DM)	1.8	1	1.3	1.2	3.4	1.7	1.3	1	1
Iron (mg/kg DM)	9.0	3.2	5.8	46	133	150	200	200	200
Copper (mg/kg DM)	2.7	1.6	9	5	40	1.6	8.0	7.3	2.8

ever possible, therefore, it is always good policy to ensure that essential minor nutrients are incorporated in a palatable, rationed element of the diet so that all animals get enough and none gets too much.

The other major mineral elements are magnesium, sodium and potassium. Sodium and potassium deficiencies are unlikely to occur except as complications of disease conditions like diarrhoea. Magnesium deficiency can occur in sucklers' calves, as indicated above and it is a possible cause of sudden death in veal calves if the diet has been improperly formulated.

TRACE ELEMENTS

The important trace elements for the calf are copper, iron, cobalt, manganese, zinc and selenium. Manganese and zinc can be ignored, as the possibility of a deficiency of either element in practical circumstances is very remote. Deficiencies of the other trace elements are covered in Chapter 6. The essential features of all these trace elements are that they are (a) required in very small quantities and (b) excreted very slowly, so they accumulate in the body. These features have three important, practical implications. It is not necessary to provide a daily supply; it may be easier and more precise to provide, say, copper or iron by intramuscular injection. Intake of, say, copper or selenium in excess of the animal's capacity to excrete them can lead to accumulations which may be toxic.

Finally, the trace element status of the calf at birth is determined very largely by the status of its mother. This acquires particular importance in the case of the veal calf which is reared on a diet low in iron, to meet the fashionable demand for 'white' veal. It is usually possible to restrict iron intake so as to produce 'white veal' without causing clinical iron deficiency anaemia (Chapter 8), but cases of anaemia do occur, especially in calves having a low iron status at birth because they were born to iron-deficient cows. The same principle can be applied to copper and probably selenium.

VITAMINS

Vitamins are usually defined as organic compounds essential in small quantities to sustain normal metabolism which cannot be synthesised by the animal and so must be provided in the diet. This definition is a little inadequate. Ruminants and other vegetarian animals do not necessarily eat the vitamin as such but may eat a so-called precursor which can be converted in the body into the true, metabolically active vitamin. Human nutritionists divide the vitamins into the fat-soluble (A, D, E and K) and the water-soluble vitamins (B complex, C). As far as ruminants are concerned, the water-soluble vitamins are not really vitamins at all, by definition. Vitamins of the B complex are synthesised by rumen micro-organisms and subsequently absorbed by the calf. Ascorbic acid (vitamin C) is manufactured by the calf itself.

The vitamins of importance to the calf are the fat-soluble vitamins A, D and E. These again are nutrients which can be stored in the body. However, the calf is born with little or no reserve of these vitamins. They are normally acquired by drinking colostrum, which is another good reason why colostrum is so important. However, the vitamin status of the cow's colostrum depends very much on her own nutrition. The cow synthesises her own vitamin A from the precursor carotene, which is plentiful in green foods like fresh grass and well conserved silage but poor in hay and roots. Synthesis of the D vitamins from their precursors, ergosterol and calciferol, requires the action of sunlight. A cow housed over winter and fed hay as her main forage is likely to secrete relatively little vitamin A in her colostrum. There is a general tendency for the vitamin A and D status of calves to decline with the winter and this is undoubtedly a major contributor to the fact that calves born in late winter tend to be more susceptible to infection than those born at the end of summer.

The classic signs of vitamin A deficiency are those of night-blindness and other forms of visual impairment. However, these signs are usually secondary. The primary role of vitamin A concerns the maintenance of the epithelial surfaces of the body, especially the gut and the respiratory tract and, in consequence, the most important practical result of vitamin A deficiency is an increased incidence of diarrhoea and pneumonia. Here again, the stockman who waits for the classic diagnostic signs of vitamin A deficiency has waited too long.

Deficiencies of vitamins D and E tend to be rather more specific and are discussed in Chapter 6.

Feed composition

Table 4 gives approximate nutritive values for some of the more important feeds for calves. These values, of course, give no indication of the variation that can exist between samples of feeds, nor of the decline in the nutritive value of grasses and grass forages with increasing maturity. More complete tables of feed composition are given by McDonald and others (1981) and by ADAS (1976). Nutritive values are expressed in Table 4 per kg dry matter. In one sense this makes things more difficult, since you have to multiply this value by the proportion of DM to obtain the nutritive value of the material (as fed). On the other hand, it is a most useful convention since it permits direct comparison between feeds differing widely in DM, e.g. hay and silage.

COLOSTRUM

Colostrum or 'beestings' is essentially milk reinforced with plasma proteins and vitamins, which concentrate in the udder in the last few days before calving (Table 4). The plasma proteins consist mainly of the lactoglobulins which transfer passive immunity from mother to offspring. The protein concentration of colostrum is about 150 g/kg immediately after birth but falls rapidly with successive feeds (or milkings) so that by the end of the second day it is little different from that in normal milk. The concentration of vitamins A, D and E in colostrum is about five times greater than that in milk although, as indicated earlier, this depends very much on the vitamin status of the cow.

MILK

Milk is usually considered as the perfect food for young calves and, on the whole, this is true on two counts: it is balanced to meet the nutrient requirements of the very young animal and it is easily digested. However, it cannot sustain normal growth on its own, being deficient in magnesium, iron and copper. The digestible carbohydrate in milk is lactose. The protein is made up principally of casein (26 g/kg) which, together with milk fat, forms the curd in the abomasum, and secondly of whey protein (10 g/kg) which is a mixture of albumins and globulins. Milk fat exists in globules about 3–4 μm in diameter and this size appears to be critical for proper digestion. It is certainly necessary to ensure that added fats in milk replacer diets are treated to give a particle size close to this. Fat globules over 10 μm in diameter may lead to diarrhoea and occasionally to alopecia – patchy hair loss which probably occurs when faeces containing improperly digested fats come in contact with the skin. The composition of milk replacer diets is discussed in Chapter 3.

BARLEY

In the U.K., barley is the most popular of the high energy grains for incorporation into animal feeds. It is rich in ME and, for a cereal, reasonably good as a source of protein. The amino acid composition of the protein is not ideal for young animals, being deficient especially in lysine. Efforts have been made to develop high protein, high lysine strains of barley but, so far, these have given rather disappointing yields. At present it is still best to consider barley as a good source of energy but one which requires protein supplementation, especially for young animals. Barley is a reasonably good source of phosphorus and magnesium but is deficient in calcium. It is advisable to break down the husk of barley by rolling, bruising or by chemical treatment before feeding it to cattle, since they have a habit of swallowing the grains without chewing them. Whole grains often pass rapidly through the rumen and can appear in the faeces remarkably unaffected by their passage through the gut.

MAIZE (CORN)

The mature corn grain is an excellent scource of energy, consisting mainly of starch, with very little fibre. Uncooked maize is less rapidly fermented in the rumen than barley and a significant amount of starch can enter the abomasum and be subjected to acid digestion. Its mineral content is similar to that of barley. It is, however, poorer in protein with respect both to quantity and to quality, being very deficient in several amino acids (especially lysine and tryptophan). The main role of maize grains in calf feeding is as cooked, flaked maize which is rightly a highly popular constituent of the coarse mix type of calf starter diet used by stockmen who wish to encourage calves onto solid food as soon as possible. It seems the young calf is very partial to corn flakes with its milk.

SOYA BEAN MEAL

This is the residue resulting from the solvent extraction from the soya bean of oil for human consumption. The residue is rich in protein (505 g/kg), having a very satisfactory amino acid composition. It is also rich in energy, phosphorus and many of the trace elements and is a thoroughly useful material for incorporation into compound feeds. However, the soya bean, in common with many of the high protein oil seeds, contains a number of more or less poisonous substances. These include protease inhibitors and substances which can cause food allergies. The presence or absence of these toxic substances in soya bean meal depends very much on the industrial processes used first to extract the oil and subsequently to treat the meal. When soya bean meal is fed to adult ruminants, most of the protein, including the toxic substances, is degraded by microbial action in the rumen, so the preparation of the meal is not critical. However, if soya bean meal is to be included in calf starter diets and, more especially, if soya flour is to be included in liquid milk replacer diets, it is absolutely essential to ensure that the toxins, and especially the food antigens, have been destroyed (Chapter 3).

PASTURE GRASS

The nutritive value of grass depends very much on its stage of maturity. As the ratio of stem to leaf increases, so MAD fibre increases, ME declines and protein declines. The ME value of grasses and grass forages can be estimated from the following equation (which is not applicable to compound feeds):

$$\text{ME (MJ/kg DM)} = 14.0 - 0.014 \text{ MAD fibre (g/kg)}$$

The element of uncertainty attached to this equation is pretty vast. On the whole, spring grass, which is growing rapidly and is rich in sugars, tends to have a higher energy value than equally leafy autumn grass having the same MAD fibre content. Even at its best, however, pasture grass falls some way short of being a complete feed for calves less than six months of age. The calf at pasture with its mother can balance the deficiencies in grass by frequent drinks of milk and thus sustain a very acceptable growth rate. I have seen cases, however, where calves of 2–3 months have been kept on good grass (or excellent silage, which is similar) and given restricted access to their mothers once or twice a day. In these circumstances they do not get enough milk to supplement the grass and growth rates have been poor or non-existent.

The mineral content of grass varies according to soil type, season, fertiliser treatment, etc. In the summer, grass is usually adequate and nicely balanced with respect to calcium and phosphorus. Magnesium levels tend to be low in spring and this problem is made worse by application of fertiliser (especially potash). (However, the problem of acute hypomagnesaemia is much too complicated to be considered simply in terms of the magnesium content of the pasture.) Dormant winter grass tends to be deficient in all the major minerals. Regional differences in trace elements such as copper, cobalt and selenium are enormous and quite outside the scope of this book (see Underwood,

1980). It is a useful general rule, however, to assume that grass, on its own, is not balanced with respect to minerals and trace elements for young, fast growing cattle and therefore that some supplementation will be necessary. It is also useful to remember that free-choice mineral blocks are not a very successful way of providing that supplement.

HAY AND SILAGE

Silage has rightly replaced hay as the winter forage of choice for highly productive ruminants. This is because it is cut when the grass is less mature and less of the original nutrients are lost during correct ensiling than during hay-making. Thus good silage is very nearly as good as good grass. Poorer silage may still appear quite attractive to a nutrition chemist, but cattle are less easy to convince. It may be fairly rich in nutrients, but that counts for little if cattle refuse to eat their appointed ration. The mineral content of silage is, on the whole, similar to that of grass. However, in wet weather, or on rough ground (or ground studded with molehills!) it is possible to pick up a lot of soil during silage harvesting and these minerals tend to get into the cattle who eat silage fairly indiscriminately.

Not a lot of silage is fed to young calves. I think the main reason for this is the fact that silage tends to be much more difficult to handle than hay in the average calf-rearing shed. There is, however, a fairly general belief among stockmen that good, soft meadow hay is somehow a 'better' feed for young calves than silage, despite the fact that the nutritive value of silage is clearly superior (Table 4). This could be because good hay is more palatable than good silage. I am not convinced of this, since I have seen calves acquire an avid appetite for silage by 3 weeks of age. It could be that hay, despite its lower nutritive value, stimulates the rumen microbial population to develop faster than does silage. This is not proven, but there is some evidence to this effect (A.R.C., 1980). It is, however, common to see ulcers in the abomasum of calves less than six months of age (Chapter 3). It is not known for sure what causes these but it is reasonable to assume that undigested particles of hay or straw which are known to enter the abomasum of young calves would be more likely to irritate and extend these ulcers than fresh grass or silage.

BARLEY STRAW

This has practically no nutritive value for the young calf and is probably best thought of as bedding. The value of barley straw as a source of ME can be improved to about 8 MJ/kg DM by treatment with alkalis (e.g. sodium hydroxide) which partially disrupt the structural carbohydrates. Even then, it contains no other nutrients of consequence. The fact remains, however, that calves will eat quite a lot of barley straw if it is their sole source of roughage. If they are to be reared intensively for beef, particularly if they are subsequently to be fed predominantly on concentrates (the 'barley-beef' system), then inclusion of unchopped barley straw to about 15% of the diet can be positively beneficial by slowing down fermentation and encouraging normal rumen function. Veal calves bedded on barley straw eat about 200 g/d, for want of any other solid food. This appears to provide some oral satisfaction but otherwise has little, if any, effect on growth or the utilisation of milk (Chapter 8).

Development of digestive function

I described the very beautiful process of development of the digestive tract from that of the newborn animal to that of the fully mature ruminant very briefly in Chapter 1. It is necessary now to examine this process in more detail.

In a newborn Friesian calf the capacity of the abomasum is about 1.5 litres and that of the reticulo-rumen about 1 litre. This means, for a start, that if any calf less than a week old spontaneously consumes (or is drenched with) more than 1.5 ℓ of fluid, it will overflow the abomasum. Secretion of the gastric, pancreatic and intestinal juices is very undeveloped at birth. In particular, the newborn calf has very few active parietal cells in the wall of the abomasum, although their numbers increase tenfold in the first few days of life. The parietal cells secrete hydrochloric acid (HCl), which is responsible for the very high acidity (or low pH) of the contents of the developed abomasum. The low acidity (or neutral pH) of the neonatal abomasum is, on the one hand, a good thing, since it ensures that immune proteins in colostrum are not digested en route through the abomasum and can thus be absorbed through the epithelium of the neonatal small intestine in the precise structural form essential to their role as antibodies and other defenders against infection. However, the low acidity of the neonatal abomasum also constitutes a potential risk to the animal, for bacteria (and probably viruses) taken in through the mouth will not be killed by acid digestion either, and can pass through to the small intestine where they are most likely to do harm. It is inevitable that calves will pick up bacteria in the first days of life. Indeed, it is essential that they do so to set off the normal process of rumen development. However, the first bacteria to colonise the digestive tract include *Escherichia coli* and *Clostridium perfringens* which are potentially capable of causing enteritis or enterotoxaemia. In the normal calf, however, enteritis does not ensue because the bacterial antigens are present in a controlled form under effective cover of maternal antibodies acquired through the colostrum. This illustrates the important general maxim that enteric diseases in calves are not simply the result of picking up an infection. These diseases, like many others, depend on a subtle balance between the infective organisms and the animal's defence mechanisms. Infection may be considered to be the natural state, and disease simply an expression of the fact that infection and resistance have got out of equilibrium.

The intestinal epithelium is effectively closed to the passage of large molecules like antibodies by about 36 h after birth. Closure appears to be governed in part by the presence of food in the gut, so if calves have drunk nothing, the epithelium stays open a little longer. However, for practical purposes it can be assumed that colostrum given to calves over 2 days old will not contribute antibodies circulating in the blood and lymph, though it may have some local protective effect against infection, perhaps reducing attachment of bacteria to the intestinal wall.

DIGESTION OF MILK

By about 6 days of age the normal processes of acid digestion have matured considerably (Fig. 3). Milk entering the abomasum via the oesophageal groove is coagulated into the curd and whey within 3–6 min. In the very young calf, rennin is the enzyme principally responsible for forming the curd. The precursor, prorennin, is secreted from the chief cells in the epithelium and rapidly converted to rennin in the abomasum. Rennin coagulates milk fastest at a pH of 6.5, which is almost neutral. As the calf develops, the parietal cells secrete more and more HCl and the chief cells begin to secrete pepsinogen which is converted to pepsin in the abomasum. Pepsin also rapidly coagulates milk but at a lower pH (5.2 is optimal). Both pepsin and rennin break down the main milk protein, casein, inside the clot. The essential difference between the two is that pepsin can break down most proteins whereas rennin is specific

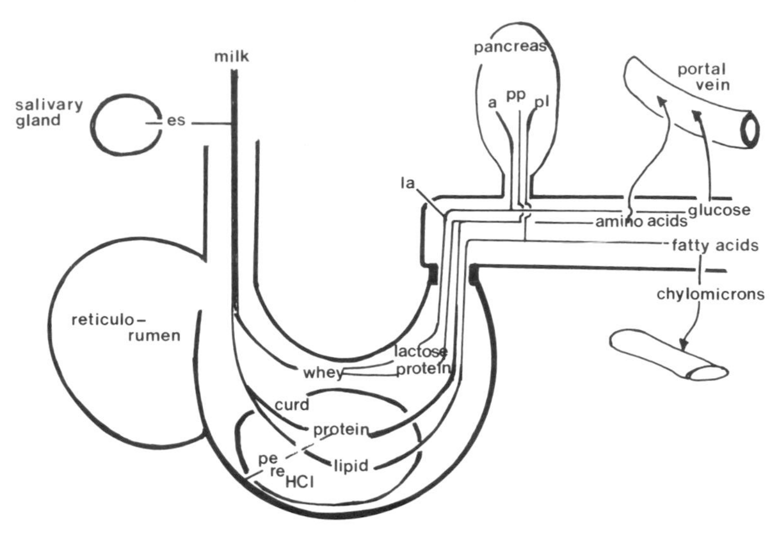

Fig. 3. Processes of digestion and absorption in the 7-day old calf drinking only milk. Key: a = amylase, es = salivary esterase, la = lactase, pe = pepsin, pp = pancreatic protease, pl = pancreatic lipase, re = rennin.

to casein. Until the HCl/pepsin system of protein digestion has developed, casein from cow's milk, or a milk replacer based on skimmed cow's milk, is the only protein that can be digested properly in the abomasum. Pepsin digestion is quite efficient by 7 days of age and there are other proteins in pancreatic and duodenal secretions which normally break down whey proteins, but it is fair to say that, because of the specificity of the rennin mechanism, no protein substitute for casein can ever quite compare with it for digestibility and therefore for safety in the very young calf.

The abomasal clot entraps both casein and milk fat which starts to break down into its fatty acids under the action of an esterase secreted in the saliva. The clot releases partially degraded proteins and fats very slowly over the first six hours after a meal and then starts to break up. When calves drink a second meal within 6 hours of the first, a new clot forms round remnants of the old one.

The whey, containing some proteins, minerals and the carbohydrate lactose, passes fairly rapidly out into the duodenum. The proteins are degraded mostly by proteases from the pancreas. Lactose is broken down to glucose and galactose by lactase. Again, this enzyme is specific to lactose. The pancreatic and duodenal enzymes of the very young calf have almost no ability to deal with starches. This ability develops with age as starch is gradually introduced to the diet.

DEVELOPMENT OF THE RUMEN

Within a few hours of life the calf will begin to investigate the environment with its mouth and start to pick up solid food and micro-organisms along with it. These pass into the reticulo-rumen and the process of microbial fermentation begins. Among the first micro-organisms to dominate in the rumen are species of *Lactobacillus* (which ferments milk, amongst other things) and *E.coli*. These are organisms which do not necessarily require oxygen but can live in the presence of it. Rumen fermentation is, however, an anaerobic process, i.e. it takes place in the absence of oxygen, and most of the micro-organisms of the mature ruminant are those for which an anaerobic environment is highly desirable or indeed essential. Among the large range of species which make up the normal microbial population of the mature rumen, one may cite the following: *Bacterioids, Ruminococcus* and *Selenomonas*, which are collectively responsible for fermenting starches and sugars, and *Methanobacterium* which converts hydrogen resulting from anaerobic fermentation into methane. As these populations develop they compete with, and may eventually eliminate from the rumen, species like *E.coli, Lactobacillus* and even outright pathogens like *Salmonella*. The normal micro-organisms of the rumen therefore constitute the most important deterrent to bacteria which have the potential to cause enteritis. This is one good reason for getting calves onto a reasonable amount of solid food as soon as possible.

The increasing intake of solid food, and particularly the end products of rumen fermentation, which are principally the steam-volatile fatty acids (VFA), acetic acid, propionic acid and butyric acid, stimulate rapid rumen development. The volume of the reticulo-rumen enlarges so that in a weaned calf of about 3 months of age it constitutes about 85% of the total volume of the four stomachs. Moreover, the papillae on the surface of the ruminal epithelium enlarge in order to increase the surface area available for absorption. The extent of papillary development is extremely dependent on the nature of the fermentation process.

DIGESTION IN THE RUMEN

The grazing ruminant receives most of its nutrients directly from the process of microbial fermentation in the rumen. This process is illustrated in Fig. 4. The animal eats solid food. During the process of eating, and more especially during the process of rumination, the food is chewed, which helps to break up the cell walls, and mixed with saliva, the most important constituents of which are sodium bicarbonate, a buffer which helps to maintain a stable pH in the rumen, and urea, which can act as a source of N for microbial protein synthesis. The carbohydrates, starch, sugar, cellulose and hemicellulose, are fermented under anaerobic conditions by rumen micro-organisms. They do this to provide themselves with energy for their own maintenance, growth and reproduction. Fortunately for the host animal, however, the micro-organisms can only capture about 6% of fermented energy for their own purposes. Some is lost as heat and methane but about 75–80% of fermented energy is excreted by the rumen micro-organisms in the form of the volatile fatty acids (VFA), principally acetic acid, propionic acid and butyric acid or their salts, acetate, propionate and butyrate. These are absorbed across the rumen epithelium and act as the major source of energy for any ruminant (Blaxter, 1968). All these three VFA

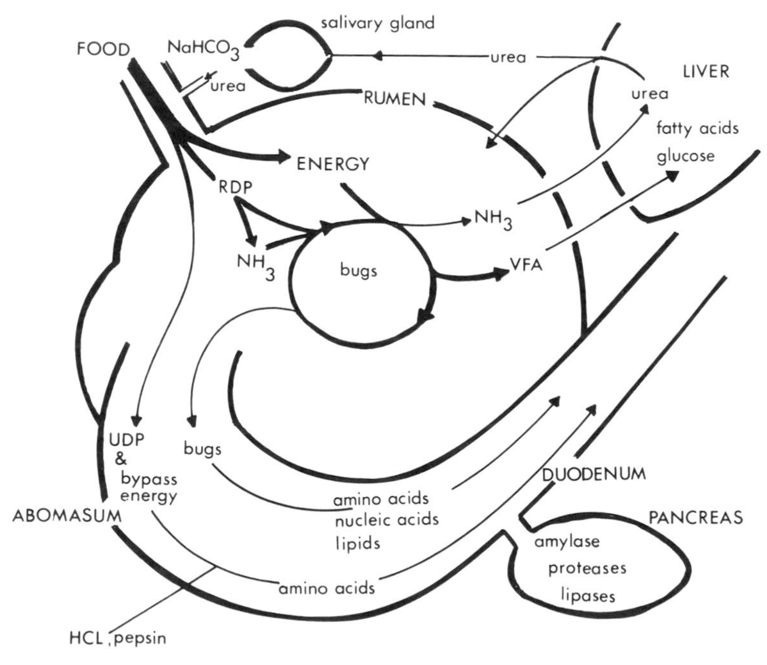

Fig. 4. Digestion in the rumen. Key: RDP = rumen degradable protein, UDP = undegradable dietary protein, VFA = volatile fatty acids.

are used with similar efficiency by the ruminant, provided all other nutrients are in balance. It is important, however, to know that propionate is the only one of the three major VFA that can give rise to a net synthesis of glucose. All ruminants have an absolute requirement for glucose. Normally the balance between acetate and propionate is such that the ruminant can manufacture enough glucose for its needs from most conventional feeds. However, when glucose requirement is elevated, e.g. during late pregnancy or early lactation, propionate synthesis in the rumen can be increased by increasing the ratio of starch to cellulose in the diet (e.g. substituting barley for hay).

The rumen micro-organisms obtain most of their nitrogen for protein synthesis from ammonia. This is produced by organisms which break down dietary protein, or salivary urea, or dietary sources of protein nitrogen (again usually urea) to ammonia, which is then taken up by other bacteria and incorporated into microbial protein. This is carried forward into the abomasum, digested, and the amino acids absorbed to meet the animal's protein requirement. The extent to which dietary true protein is broken down by microbial action in the rumen depends on (a) its vulnerability to microbial attack, and (b) the length of time it spends in the rumen. Dietary crude protein is now described in terms of two constituents: *rumen degradable protein* (RDP) and *undegradable dietary protein* (UDP). RDP, which may include true protein and NPN, is broken down in the rumen and incorporated into micro-organisms, mostly via ammonia, but in part as amino acids, at a rate determined by the energy metabolism of the organisms and therefore their rate of VFA synthesis (Ørskov, 1982). It follows therefore that the rate at which the rumen micro-organisms can provide protein to the host animal (as microbial protein) is fixed within rather narrow limits by the rate at which they can provide energy as VFA. Making a lot of assumptions which need not concern us here, the Agricultural Research Council (1980) have proposed that microbial protein N can be related to ME generated by ruminant fermentation and digestion according to the formula

7.8 g microbial crude protein = 1 MJ metabolisable energy

If the ratio of RDP:ME exceeds this, the excess N will be converted to ammonia but not captured by the rumen micro-organisms, and so wasted. The young, rapidly growing calf has a requirement for dietary protein relative to ME that exceeds this ratio of 7.8:1. It is important therefore to ensure that a sufficient proportion of dietary protein is in the form of undegraded dietary protein (UDP) which is that portion of the dietary true protein which escapes fermentation in the rumen and can pass on to the abomasum for acid digestion. In most vegetable feeds the protein is highly degradable (over 70%), e.g. barley, soya, groundnut meal, etc. Animal protein sources, e.g. fish meal, contain a relatively high proportion of UDP (degradability less than 40%). There are now ways of treating protein sources like soya bean meal to reduce their overall degradability. These include certain forms of heat treatment and the use of formaldehyde.

To recapitulate: the importance of UDP lies in the fact that the ratio of protein to energy *supplied* by the rumen micro-organisms is fixed, whereas the ratio of protein to energy *required* by the animal increases with increased level of production (growth and lactation). The young, rapidly growing calf requires a ratio of protein to energy considerably greater than that which the rumen micro-organisms can provide. When it is eating grass and sucking milk from its mother, high quality milk protein bypasses the rumen through the elegant mechanism of the oesophageal groove. If the calf has been weaned at, say, five weeks of age onto solid food, this mechanism is not available and, to ensure optimal growth, it is necessary to ensure that a significant amount of dietary protein is in such a form that it escapes degradation in the rumen (see Table 5).

Nutrient requirements

ENERGY

The energy requirements of any animal are arbitrarily divided into requirements for maintenance and growth. The maintenance requirement of a calf is determined mainly by the animal's size, to a lesser extent by its level of activity and (rarely) by the need to combat the stress of cold (Chapter 4). ME requirement for growth depends on the rate of weight gain desired, the composition of the weight gain and the net efficiency of utilisation of ME for growth (A.R.C., 1980).

The ME requirement of any warm-blooded animal can, as a first approximation, be related to body weight using the classic concept of metabolic body size. Kleiber (1961) showed that the maintenance energy requirement of all birds and mammals was an approximately constant function of body weight raised to the power 3/4. This law expresses the fact that the amount of ME

Table 5 Requirements of growing calves for ME, RDP and UDP

	Liveweight (kg) 80	140	200
Maximum DM intake (kg/d)	2.4	3.6	4.8
ME requirement (MJ/d)			
maintenance	15	23	30
maintenance + gain (kg/d) 0.25	18	27	36
0.5	22	32	42
0.75	26	38	48
1.0	31	43	55
Crude protein requirement (g/d)			
0.5 kg/d gain RDP	170	250	330
UDP	130	120	110
1.0 kg/d gain RDP	240	335	430
UDP	200	180	150
Minimum crude protein in DM (g/kg)			
0.5 kg/d gain	125	100	90
1.0 kg/d gain	185	140	120
Optimum degradability of protein (%)			
0.5 kg/d gain	56	68	75
1.0 kg/d gain	55	65	74

required *per kg bodyweight* declines in a curvilinear fashion with increasing body weight, whether the comparison is made within or across species. Using logarithms to straighten the curve,

$$\log \text{ME}_{\text{maintenance}} = a \times 0.75 \log \text{bodyweight (kg)}$$

Applying this law to the growing calf, and erring slightly on the side of generosity to take account of things like activity, one can predict ME requirement for maintenance thus:

$$\text{ME}_{\text{maintenance}} = 0.55 \times \text{bodyweight (kg)}^{0.75} \text{ (MJ/day)}$$

To take one example from Table 5: the *metabolic body size* of an 80 kg calf is 26.7 kg ($80^{0.75}$). Multiplying this by 0.55 gives 14.7 MJ/d, which rounds up to 15 MJ/d. The first advantage of this general law is that one can reasonably accurately predict the ME requirements for maintenance of any calf of any size or age. The second advantage is that one can equally easily predict ME requirements for satisfactory growth since most calves fed conventional diets of grass and cake will consume and grow most efficiently at an ME intake close to, or a little above, twice that required for maintenance.

Table 5, which is adapted from A.R.C. (1980) gives examples of the ME requirements for calves to sustain different rates of gain at different body weights. Note that in each case a liveweight gain of 1 kg/day corresponds to an ME intake of twice maintenance or slightly less. It is obviously advisable not to push too much feed into calves shortly after weaning and at this stage, restriction to twice maintenance or below is sensible. If you want to achieve maximum rates of gain in calves of 150 kg or over, then aim for an ME intake of about 2–2.5 times maintenance. There are quite small differences between breeds and between feeds in ME requirement for growth and these are considered by A.R.C. (1980).

The appetite of a calf is limited in part by its energy requirement but also by the capacity of the gut. There are, of course, enormous differences between individuals in appetite, but, again as a first approximation, one can relate DM intake of solid foods to metabolic body size, it being about 90 g DM/kg weight$^{0.75}$ per day. This corresponds to a DM intake of 2.4 kg/d at 80 kg and 4.8 kg/d at 200 kg liveweight (see Table 4). To achieve the ME requirement for maintenance plus 1 kg/d liveweight gain at 200 kg, and within the limits of appetite, would require a mixture of feeds having an overall ME content in the DM (M/D) of 11.5 MJ/kg DM (55/4.8). Given a compound feed with an M/D of 13 and hay with M/D of 8.6, the calf would require

3 kg DM compound feed @ 13 M/D	=	39 MJ
Hay (*ad lib.*) = 1.8 kg DM @ 8.6 M/D	=	15
Total	=	54 MJ/d

Such a ration would be perfectly acceptable to the calf but perhaps too expensive. Suppose the farmer was prepared, at this stage of growth, to accept 0.5 kg/d liveweight gain. The ration could then be reformulated

0.8 kg DM compound feed @ 13 M/D	=	10
4.0 kg DM hay (*ad lib.*) @ 8.6 M/D	=	34
Total	=	44 MJ/d

The old stockman's maxim of 'two pounds a day of cake' corresponds almost exactly to 0.8 kg DM/d. It is fair to say that this is about the lowest level of addition of compound feed compatible with reasonable growth in calves under six months of age.

PROTEIN

The requirements of the growing calf for RDP and UDP can be related to ME requirement (Ørskov, 1982). As yet, however, it is not possible to predict protein requirements with the same precision as has been achieved for energy requirements, so the protein values given in Table 5 tend to err, if anything, on the side of generosity. Requirements for RDP and UDP are given at rates of gain of 0.5 and 1 kg/d. These values are my best estimates of what a typical Hereford x Friesian steer calf would require; bulls and the large exotic breeds would require 10–15% more UDP. Table 5 also converts these values into more practical units, namely the minimum crude protein for the DM of the complete ration and the optimal degradability (RDP:CP), expressed in percentage units. It shows how the CP content of the DM has to increase with increased rate of gain. At 200 kg and over, the optimal degradability of dietary CP is about 75%. This is the extent of degradation in the rumen one would expect for a conventional diet based on hay, barley and soya or groundnut cake.

Below 200 kg, or perhaps 300 kg for bulls of the large exotic breeds, the optimal degradability is 65% or lower, so the compound feed should include sources of UDP in order to bring its overall degradability down to about 55%.

MINERALS AND VITAMINS

Requirement of growing calves for the major minerals (A.R.C., 1980) are shown in Table 6, both as absolute daily requirement and as minimal concentrations in the DM of the diet. These latter figures can be compared with those of Table 4 which lists the mineral contents of feeds. Note that pasture grass, for example, would provide adequate Ca and P for 0.5 kg liveweight gain per day in a 200 kg calf. At 1.0 kg/d gain, the grass would be marginal for Ca and slightly deficient for P. Note also how the cereal grains are deficient in Ca.

Requirements for the trace elements and fat soluble vitamins are also given in A.R.C. (1980). I shall not reproduce them here since, for reasons stated already, it is not very helpful to think in terms of a daily requirement for these accumulated minor nutrients.

Table 6 Requirements of growing calves for the major minerals (from A. R. C. 1980)

		Liveweight (kg)			
		100		200	
	Daily gain (kg)	Daily requirement	Content in DM	Daily requirement	Content in DM
Calcium	0.5	12 g	4.2 g/kg	14 g	3 g/kg
	1.0	21 g	7.5 g/kg	24 g	5 g/kg
Phosphorus	0.5	6 g	2 g/kg	8 g	2 g/kg
	1.0	11 g	4 g/kg	13 g	3 g/kg
Magnesium	0.5	3 g	1 g/kg	4.8 g	1 g/kg
	1.0	4.2 g	2 g/kg	6 g	1.5 g/kg
Copper	0.5	31 mg	11 mg/kg	48 mg	10 mg/kg
	1.0	44 mg	16 mg/kg	61 mg	13 mg/kg

Problems of deficiencies are sporadic and will be considered along with other diseases in Chapter 6.

WATER

Water is obviously an essential element of any diet and since it is cheap it is obviously good husbandry to provide calves with as much fresh, clean water as they want. When calves are on all, or nearly all liquid diets, having a DM content of about 125 g/kg, their water intake exceeds their physiological water requirement and they will not be stressed by the absence of extra water unless exposed to heat stress (Chapter 4). This is not to say that one should deny them access to it. As soon as calves start to eat significant amounts of solid food, especially dry forages like hay, they require continuous or regular access to fresh water. Indeed, it has been shown that calves which were bucket-fed milk replacer once or twice a day prior to weaning increased their consumption of hay at a much greater rate when fresh water was always available between liquid feeds. This observation is of particular interest to rearers of heifers for dairy herd replacement, since one of the most economically important characteristics of a good dairy cow is a large appetite for grass and forage.

Further reading

Agricultural Development & Advisory Service (1976) *Nutrient Allowances and Composition of Feedingstuffs for Ruminants.* ADAS Booklet 2087 MAFF, London.

Agricultural Research Council (1980) *The Nutrient Requirements of Farm Livestock. No. 2 Ruminants.* 2nd ed. Agricultural Research Council, London.

Blaxter, K.L. (1967) *The Energy Metabolism of Ruminants.* Hutchinson, London.

Kleiber, M. (1961) *The Fire of Life.* Wiley, New York.

McDonald, P., Edwards, R.A. & Greenhalgh, J.F.D. (1981) *Animal Nutrition.* Oliver & Boyd, Edinburgh, 3rd ed.

Ørskov, E.R. (1981) *Protein Nutrition in Ruminants.* Academic Press, London.

Underwood, E.J. (1980) *Mineral Nutrition of Livestock.* Commonwealth Agricultural Bureaux, Slough.

3 Feeding systems from birth to weaning

This chapter is concerned with feeding the calf born to the dairy cow from the time it is taken from its mother, or ceases to be fed colostrum, until the time it is weaned onto a diet of solid food. At first, the calf is given only a liquid diet of milk or a reconstituted milk replacer powder having a composition very close to that of cow's milk. Within the second week of life it is introduced to a forage, usually hay, and a 'starter' ration based on cereals and protein-rich meals supplemented with minerals and vitamins. Weaning off milk replacer takes place when the calf is consuming sufficient solid food to sustain maintenance and a reasonable rate of growth.

Different fashions in calf husbandry in different areas of the world have tended to favour different ages as optimal for weaning dairy calves, from as early as four to as late as twelve weeks after birth. However, the trend over the last 20 years has been towards weaning as early as possible. There are several good reasons for this:

(1) Milk replacer costs about three times as much as calf starter pellets per unit of metabolisable energy.
(2) Feeding milk replacer to calves is time-consuming, particularly if calves are kept in individual pens and fed from buckets two (or even three) times per day.
(3) Pens for rearing calves from birth to weaning are often in rather high-cost accommodation and obviously the shorter the period to weaning, the fewer pens required.
(4) There is a fairly widespread belief that diarrhoea ('scours') is really only a problem before weaning.

Taken together, all these factors make up a strong case for weaning calves as soon as possible and getting them out into group pens in relatively low-cost, naturally ventilated 'follow-on' units. However, no single factor, taken on its own, is sufficient reason for weaning calves at the earliest possible time, which is about 5 weeks. A calf, bucket-fed milk twice daily to weaning at five weeks, would consume about 16 kg of milk powder. This represents about 0.5% of the ME required to bring a beef animal to slaughter weight or a dairy heifer to the point of first calving. Even if it were reared on a system that provided milk replacer free-choice from a teat and drank 2–3 times as much, the cost of that milk replacer would still be very small compared to total feed costs during the rearing period. Moreover, the fact that some calves scour when given milk replacer is not, by itself, sufficient reason to advocate very early weaning. The most important practical consideration in the pre-weaning period is to ensure first that the calf stays alive and second, that it does not succumb to disease severe enough to set back its growth. The most economic feeding system prior to weaning is not necessarily therefore the one that costs the least for food and labour.

The last few years have seen a large number of new ideas for feeding systems being tried out on young calves. This has come about first as a result of the increased concern to ensure successful rearing, due largely to the increase in the value of calves, and second (and for the same reason) from improvements in technology relating to calf feeds and feeding machines. I should perhaps begin discussion of these different systems by stating my own personal opinion, which is that there is no one 'best' system for rearing calves to weaning. I have seen successes and failures with them all.

I shall first describe the various systems for feeding calves from birth to weaning, pointing out only their more obvious advantages and disadvantages. Then, having also described the composition of milk replacers and starter feeds, I shall discuss the various feeding systems in the context of physiology, production and economics, health and welfare. It is ultimately up to the stockman, with advice from his own feed supplier and veterinary surgeon, to decide what system best suits the unique circumstances of his own unit.

Bucket-feeding systems

In most bucket-feeding systems calves are kept in individual pens, each with two buckets, one for milk replacer and water, the other for the calf starter ration. Pen design is considered in Chapter 4. There are differing and rather strongly held beliefs about the best time at which to remove the calf from its mother and transfer it into an individual rearing pen. In most cases the calf will drink much more colostrum if kept with its mother for the first 24 hours than if it is removed shortly after birth and fed twice or three times with colostrum from a bucket. Obviously, the more colostrum it drinks the better. It is important, however, to ensure that if the calf is left with its mother for the first 24 hours, it is drinking regularly. The high-yielding dairy cow will also need to be milked at least once during the first day, as she will produce more colostrum than the calf could possibly consume.

Some dairy farmers, particularly in the south-west of England, still like to let their calves drink from their mothers for the first two weeks of life. This undoubtedly gives them a very good start and puts a 'bloom' on the animal, a shining healthy appearance to the coat, which is worth several pounds in the sale ring. As indicated earlier, cow's milk probably continues to confer some immunity against enteric infection on the surface of the gut after the epithelium has closed to the absorption of colostral antibodies. All this is to the good, although it is, of course, not a least-cost option. However, the cow–calf bond has become very strong by this time and it is very distressing to all parties concerned to separate mother and calf after about two weeks together, as anybody who has tried to sleep in the vicinity will readily witness. It undoubtedly causes less distress to separate mother and calf as soon after birth as possible.

The next problem is to train the young calf to drink from a bucket. The normal procedure is to allow the calf to suck from your finger, then lower your hand into the bucket which is inclined at such an angle that the calf's mouth, but not its nose, dips into the milk. Most calves readily pick up the idea, so that by the second or third feed they will drink from the bucket without this form of encouragement. I should perhaps point out here that the natural behaviour of the calf at the beginning of a meal is to bunt (or butt) forward with its nose to encourage the let-down of milk from the mammary gland. It pays, therefore, to hold on tight to the bucket! There is a common belief, which I share, that pure-bred Friesian calves are easier to train to buckets than calves from beef sires. For the calf that is difficult to train, you can purchase a trainer teat which can be lowered into the milk. For the rare calf that is extremely difficult, I can only counsel patience. As a very young man I was once driven nearly to distraction by a superb looking Hereford x Friesian calf that refused to drink anything (including colostrum) and fought all attempts to assist it for four days, after which it suddenly appeared to decide it was hungry, drank from a bucket unassisted and never looked back thereafter. These days I would have ensured that it received at least 4 feeds of colostrum (each 1.5 litres) by stomach tube (Chapter 6) but I would not have worried so much thereafter if it took several sessions before it consumed its proper ration.

During the first week of life, calves are sometimes given three bucket-feeds a day, but if these are properly spaced out this makes for a long, hard day and there is no good evidence to show that it is necessary to feed calves more than twice a day, even at this early age. Remember, the curd in the abomasum persists for about 8 hours. I do however believe that home-reared calves benefit from getting milk rather than milk replacer for about the first week of

Table 7 Feeding recommendations for home-reared, bucket-fed calves

Twice-daily feeding, milk replacer	
Day 1	Colostrum from cow
Days 2–3	Colostrum/whole milk; 1.5 *l* (2.5 pints) twice daily, 38–40 C
Days 4–7	Whole milk or milk replacer (125 g powder/*l*); 1.5 *l* twice daily
Week 2	Increase milk replacer to 2–2.5 *l*/feed (125 g powder/*l*); introduce calf starter ration (100–200 g at first) and hay
Weeks 3–6	Maintain milk replacer at 2–2.5 *l*/feed twice daily; provide water after each milk feed Increase calf starter ration according to appetite and provide hay to appetite Wean abruptly when intake of starter ration reaches about 1 kg/d
Once-daily feeding, milk replacer	
As above for first two weeks	
Weeks 3–5	Reduce milk replacer to once daily, increase (over 4 days) ration to 3–3.5 litres containing 450–500 g powder Wean abruptly when intake of starter ration reaches about 1 kg/d
Twice-daily feeding, milk and gruel	
Week 1	Colostrum and whole milk as above
Week 2	2–2.5 *l* whole milk twice daily, introduce starter and hay
Weeks 3–6	Gradually substitute gruel for milk, beginning with 1.75 *l* (3 pints) milk + 0.5 *l* (1 pint) gruel (at 100 g/l) and achieving 2–2.5 *l*/feed (at 100 g gruel/*l*) twice daily. Provide about 500 g/d calf starter
Weeks 7–12	Gradually reduce gruel and wean when intake of calf starter ration is about 1.5 kg/d

life. Fed at a rate of 1.5 litres twice a day, this is not an excessive drain on farm income.

It is recommended that conventional 'sweet' milk replacers based on skim-milk and added animal fat should be fed at a concentration of 125 g/ℓ and that calves should initially be given about 1.5 ℓ/feed (2.5 pints) at blood heat (38–40°C). Thereafter, the feeding schedule proceeds as illustrated in Table 7. The amount of milk replacer given to an individual calf will vary slightly (e.g. 2–2.5 ℓ/feed in week 3) according to the size of the individual calf. A starter ration in the form of pellets or a coarse mix and hay can be introduced within the second week of life. It pays, however, not to offer too much too soon, as uneaten food tends to spoil and become unpalatable. A typical arrangement for feeding calves in individual pens is illustrated in Fig. 5a.

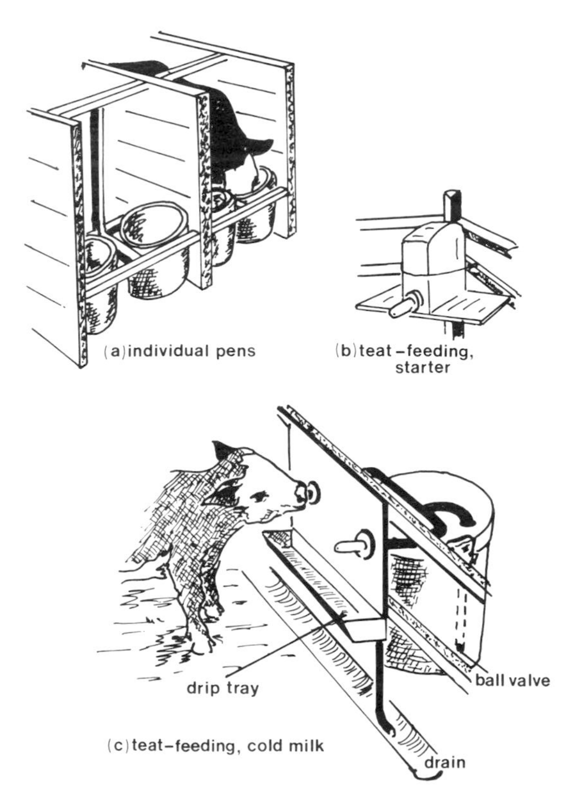

Fig. 5. Feeding arrangements for calves. (a) Individual pens with solid sides to prevent direct contact with calves, (b) Simple arrangement for starting young calves on feed, (c) Permanent arrangement for feeding cold milk to calves from teats.

One of the main objectives of rearing calves in individual pens is to ensure good hygiene. It is particularly important to ensure that this applies to all feeding utensils. Some authorities recommend that buckets be cleaned daily in hypochlorite or some other approved disinfectant. This is something of a counsel of perfection. It is undoubtedly important to keep buckets clean but the best aid to good hygiene is probably to ensure that each calf uses only its own bucket right through to weaning. Milk buckets should be rinsed out after each feed and then half-filled with fresh water, since there is clear evidence that unlimited access to fresh water helps to encourage forage intake. It is important, however, to avoid swilling water around too generously during the cleaning-out periods, as this will make it very difficult to maintain an acceptable relative humidity in the calf house (Chapter 5).

The calf that is bucket-fed twice a day is taking in 500 g/d of milk replacer by 2 weeks of age (Table 7). This provides about 9 MJ/d of ME, which is less than the maintenance requirement for a 50 kg calf. The system is clearly providing the calf with considerably less than it needs of the food it is best able to digest, and encouraging the precocious development of the microbial and enzymic processes required to digest starch and cellulose. Weaning should take place, not at any fixed age, but when the calf is eating at least 1 kg/d of starter feed (Fig. 6). Assuming this has an M/D of about 12.5 MJ/kg as fed, this too only constitutes a maintenance ration for a calf weaned at about 60 kg.

The main advantage claimed for the bucket-feeding system is that it minimises feed costs, especially the cost of milk replacer, intake of which can be restricted to about 16 kg/calf. However, for the reasons given in the last paragraph, it has carried least-cost feeding to its reasonable limit. Any calf weaned off milk replacer when eating less than 1 kg/d calf starter would expect to suffer a set-back. The other claimed advantages of bucket-feeding in individual pens are that it provides good opportunity for close inspection of individuals and early treatment when necessary and, since it ensures predictable and fairly even growth, weaning can be controlled and calves turned out in batches into 'follow-on' yards. The obvious disadvantages of this system are that it requires the most labour and tends to have the highest capital costs for pens and buildings.

The practice of once-daily bucket-feeding has evolved simply to reduce labour. It is recommended that calves be reduced from two feeds daily to one at 14 days old, the concentration of the milk

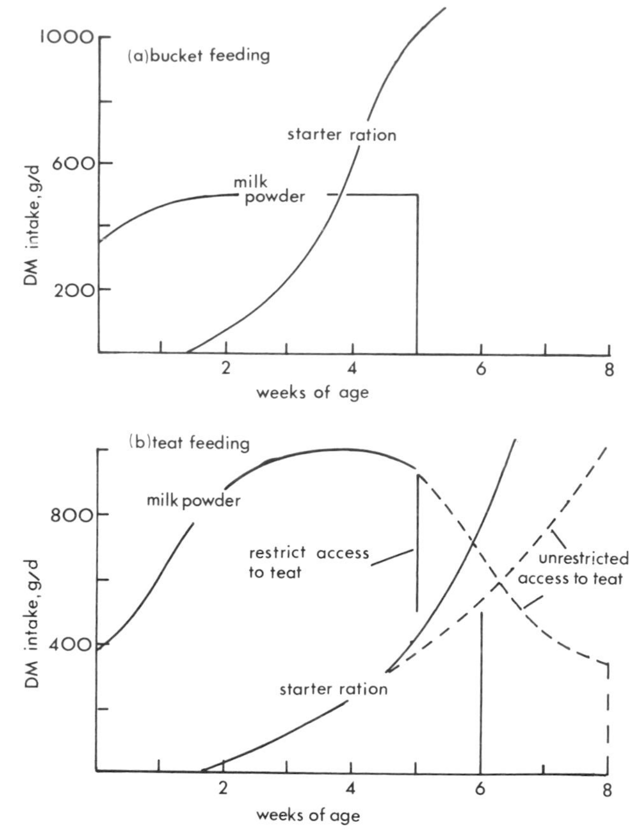

Fig. 6. Intakes of milk powder and starter rations. (a) Bucket-feeding to 5 weeks, (b) Unrestricted teat feeding to 5 weeks followed by restricted access until six weeks or unrestricted access to eight weeks.

being increased to 150 g/ℓ, providing 450 g in 3–3.5 litres (Table 7). There can be no case for recommending once-daily feeding as preferable to twice-daily feeding on grounds of physiology, health or welfare. It is no more than a labour-saving device. However, there are many farmers who operate once-daily systems extremely successfully and raise excellent calves and it would be foolish arrogance to tell them they were doing things all wrong. When the system does go wrong however, it can go very wrong indeed.

It has been traditional for home-reared calves to be put in calf boxes, often in groups of about four, and bucket-fed twice daily. This system can work very well for small groups of healthy, home-reared calves often dispersed fairly thinly about the farm. For reasons that I will develop later, it may not be good practice to bucket-rear bought-in calves in groups. When you buy in calves you can practically guarantee that you buy in infection. Preventing infection from proceeding to disease, particularly enteric disease, in the pre-weaned calf requires a rather subtle variety of skills relating to feeding management, hygiene and the use of preventive medicines. These are not really compatible with bucket-feeding infected calves in groups.

Another system that comes under the general heading of bucket-feeding is that which gradually substitutes a liquid, vegetable-based gruel for whole milk after about 2 weeks and continues to feed this from the bucket to about 12 weeks of age. Details are given in Table 7. The system is really only suitable for small groups of home-reared calves for a number of reasons. One, it usually assumes the feeding of a certain amount of whole milk. Two, it is very demanding of labour and calf accommodation. Three, liquid feeding based on vegetable starch and protein rather than lactose and casein carries a slightly higher risk of precipitating scours if calves are otherwise stressed by infection, mixing and transport. For home-reared calves however, the system is only marginally less satisfactory than bucket-feeding milk replacer up to 5 weeks of age. By 12 weeks of age it is probably marginally superior.

Teat-feeding systems

Some people who rear calves to weaning in individual pens in the manner described above prefer to allow the calves to drink their ration of milk from a teat attached to their individual bucket. In general, I see no great advantage to this though it probably pays the calf-rearer to have a few buckets with teats attached for any individual calf which may refuse to drink direct from the bucket or which may show evidence of indigestion after drinking. If a bucket with teat attached is used, it is important to arrange it so that when the calf drinks, its neck is stretched out at least horizontal or with its head tilting slightly upwards, since this position appears to aid in closing the oesophageal groove and ensuring the passage of milk replacer direct into the abomasum.

Systems of teat-feeding are usually employed when calves are reared in groups. The milk replacer can either be provided, freshly mixed and at blood heat (about 40°C) from an automatic dispenser or the calves can simply suck up cold milk replacer from a plastic bucket. Since most automatic milk dispensers for calves cost (in 1982) several hundred pounds and since one can set up a cold milk feeding system for £10 or less, it is hardly surprising that the latter system is more popular, particularly since the advent of acidified milk replacers.

The cold milk, teat-feeding system is illustrated in Figs. 5(b) and (c). It is best to use a plastic bucket, particularly if acidified milk replacer is fed. When calves are first introduced to the feeder it is useful to have the teat at the bottom of the container so that weak or confused calves can receive milk without having to suck long and hard (Fig. 5(b)). When calves are older it is better to make them suck up the milk against gravity (Fig. 5(c)), not least so that if they destroy the teat the milk will not run away. However, the milk line to the teat should contain a simple ball valve to prevent milk running back into the bucket when the calves have finished feeding. The teat needs to be strongly but smoothly secured to prevent calves pulling it off or tearing if off. We secure the rim of the teat behind a smooth metal plate which can be bolted (using butterfly nuts for quick release) to a second plate on the outside of the pen. The frame around the teat has to be big enough to prevent the calves getting their tongues round the back of the teat and pulling the milk line away. Ideally, there should also be a drip tray to catch the saliva which calves dribble profusely while sucking (Fig. 5(c)). Alternatively, the teats can be set directly above a covered drain. Since calves pass the majority of their urine and faeces in the area of the teat, it pays to ensure that this is an unbedded, well-drained area (see Chapter 5) and ideally it should be scraped clear daily.

The arrangement of teats should be the same whether cold milk is fed direct from a bucket or whether you use a machine to dispense warm milk. Whichever system you employ, it is essential to wash all the milk lines through with mild disinfectant, then with fresh water, daily. Competition between calves at the teats is not a severe problem (see Chapter 7) but the number of calves per teat should probably not exceed 10 and 4–6 calves per teat is better if this can be arranged.

The main problem with allowing calves full access to milk replacer through teats is that they prefer this, quite rightly, to any of the other food around. By four weeks of age the calves may be consuming over 1 kg/d of milk powder but eating very little hay or starter ration. Fig. 6(b) shows a typical pattern of intake of milk powder and starter on this system. If calves are weaned at five weeks when eating only about 250 g/d of starter they receive a severe setback and lose all the advantage of the good start provided by free access to milk early on. It is particularly important in this system therefore to ensure that the starter ration is as palatable as possible; a good coarse mix is especially valuable here. Again, the ideal time for weaning is when the calves are eating about 1 kg starter per day. In a group-rearing system it is impossible to tell how much each individual calf is eating. It is often the strongest and most aggressive calf which is eating the least starter because he is getting the most milk.

If weaning is delayed until about 8 weeks of age one can be reasonably sure that nearly all healthy calves are eating enough concentrate food to prevent a post-weaning setback. Free access to milk replacer to 8 weeks is not, of course, a least-cost option but it can produce some lovely strong calves. The cheaper alternative is to withdraw milk progressively for the last 14 days before the intended weaning age (Fig. 6(b)). This can be achieved (i) by restricting the amount of milk powder so that calves can suck only water for increasing periods of the night and day, (ii) by introducing a narrow bore section into the milk line so as to make it much harder for the calves to drink, (iii) by removing the teat altogether for increasing periods of time. It pays not to leave an entirely non-productive teat in the pen as the calves are likely to rip it to pieces. It is of course essential to have a separate source of fresh water for the calves to drink at this or any time.

There has recently come onto the market a range of automated calf feeding machines similar in principle to individual 'out-of-parlour' feeders for dairy cows. Attached to a collar around the calf's neck is a 'responder' or tuned-coil. This enables a microprocessor in the feeder to recognise each individual calf and make milk available to it up to its ration for a period of 24 hours or less. The double attraction of this type of machine is that it not only controls the intake of each individual but it can also provide a daily record of intake. Rationing intake cuts down on the cost of milk replacer, encourages early consumption of hay and starter and avoids the risk of calves drinking to excess and developing diarrhoea. A daily record of the intake of each calf also reveals any drop in appetite and so can act as an early warning that a calf may be sickening for some disease in the same way as does milk left undrunk in a bucket-rearing system. The cost of such a system is difficult to justify at present for conventional calf-rearing, although microcomputers keep on getting cheaper and cheaper. The automated feeder that rations each calf is, however, particularly suitable for the group-rearing of veal calves and in these circumstances could pay for itself quite quickly (Chapter 8).

Composition of milk replacers

The composition of cow's milk was described in Table 4. The three most important nutrients in dried cow's milk are the carbohydrate, lactose (360–400 g/kg), fat (300–400 g/kg) and milk protein (280–320 g/kg) made up principally of casein, which clots in the presence of rennin or pepsin and forms the curd, and the whey proteins (albumin and globulin). Skim-milk is the by-product of butter-making and consists of lactose and all the milk proteins. Whey is the by-product of cheese-making and consists of lactose and the whey proteins only.

'SWEET' MILK POWDERS (SKIM-MILK BASED)

Conventional milk replacers are made by adding animal fats to skim-milk. The technology of this process has improved markedly in recent years, particularly in regard to the emulsification of the added fats to ensure that the product mixes easily and that the diameter of the fat particles suspended in solution does not exceed 5 μm. Such milk replacers are usually marketed as 'high fat' or 'sweet' milk replacers. The expression 'high fat' is a misnomer, since their fat content is about the same as, or a little less than, the fat content of whole milk. It merely means, higher than it used to be when emulsification techniques were less developed and the fat content of skim-based milk replacers was as low as 50 g/kg. The modern 'high fat' (i.e. normal fat) milk replacer is similar in composition to whole milk and is therefore an excellent, albeit expensive feed, ideally suited to systems involving bucket-feeding warm milk replacer. The product is physiologically correct for the immature digestive system of the calf, i.e. it contains the nutrients that the calf can digest and in the right proportions. It forms a clot in the abomasum and so provides a slow release of nutrients to the duodenum. This makes it particularly suitable for once-daily feeding. It is also extremely palatable but it is, of course, liable to sour quite quickly. The expression 'sweet' milk replacer differentiates this product from the acidified milk replacers which have better keeping qualities and can be fed cold.

ACIDIFIED MILK POWDERS

The acidified milk powders come in two very distinct forms. The 'mild acid' milk powders are essentially similar in composition to the sweet milk replacers, i.e. they consist of skim-milk and emulsified animal fat, but organic acids (e.g. citric or fumaric acid) have been added to reduce the pH of the product to about 5.7. The original purpose of this exercise was to increase the keeping quality of the powder so that a mixture could be made up once daily and fed cold to calves group-reared on teat systems. It is fair to say that, except in the hottest weather, this mild degree of acidification does prevent the milk from souring within 24 hours. It is not possible to acidify skim-milk any more than this without clotting the casein and creating curds and whey in the bucket which cannot then be sucked through the teat. These mildly acidified milk powders appear to be as palatable as 'sweet' milk powders and they clot in the abomasum just as fast, if not faster. There were fears at first that *ad lib.* feeding of a cold but highly palatable milk powder would cause a serious increase in the incidence of scours but this has not proved to be the case for reasons discussed later in this chapter. Both sweet and mild acid milk powders tend these days to contain anti-bacterial 'growth promoters' like Nitrovin. These substances act only against Gram-positive bacteria which are of little importance inside the gut of the calf but which do survive in the air and are more likely therefore to contaminate milk in the bucket or powder in the bag. The expression 'growth promoter' in this context is a piece of deceptive sales talk indulged in by most suppliers and usually defended by phrases like, 'Well, our competitors include it and farmers seem to want it, so we have to go along with it'. Substances like Nitrovin added to milk powders are best thought of as preservatives designed to keep the milk fresh up to the time it is drunk.

STRONG ACID POWDERS

These are acidified to a pH of about 4.2. This immediately precludes the use of skim-milk, since casein clots at this pH. In theory, the calves could eat the curds and whey, but strong acid powders, by virtue of their excellent keeping properties, are designed for systems in which groups of calves are given free access to milk replacer dispensed through a teat. Initially, these strong acid powders were made by adding whey protein concentrate to conventional whey and fat to bring the protein concentration of the powder up to that of conventional milk powders (i.e. 230 g/kg). Milk powders containing no casein do not, of course, clot in the abomasum and there were some fears initially that these strong acid powders would give rise to a number of digestive disturbances due to the abomasum emptying too rapidly. In most circumstances, however, they have proved both safe and popular.

In the early days of strong acid powders there was a line of sales talk which claimed that they were 'formulated to reduce over-consumption'. If this had been true, it would have been quite a good selling point since over-consumption on '*ad lib.*' systems is (a) expensive, and (b) liable to induce scours. It would, however, have been simpler (if less persuasive) to say that they were apparently less palatable than conventional sweet milk powders. In fact, most calves appear to drink strong acid powders with every sigh of relish. Indeed, this degree of acidification appears to offer no protection against that scourge of the free-choice, teat-feeding system: the occasional calf which drinks itself into a stupor, gets an obvious belly-ache, scours, and then comes back for more.

A few calves reject strong acid feeds. It is particularly important therefore to be sure that all calves are drinking regularly. If an individual is clearly not taking to the system and failing to thrive, it may be necessary to remove it and rear it on conventional milk powder or whole milk from a bucket. It also seems to me that calves which are fed strong acid feeds show a slightly greater incidence of 'scalding' – loss of hair, especially around the nose, the anus and the lower limbs. It is thought that this hair loss is usually caused by contact with undigested fats in diarrhoeic faeces. Maybe this is more likely with strong acid feeds if fats pass more rapidly out of the abomasum. It may also be the case that the feed itself is sufficiently acid to scald the area around the nose. Scalding around the feet and anus can, I think, only arise from contact with wet faeces.

All in all, however, the strong acid milk replacers have to be pronounced a success. Because of their differences from whole milk, I would not, on physiological grounds, recommend that they be fed to calves less than two weeks of age, because of the immaturity of the calf's digestive system, or used in a once-daily, bucket-feeding system, because they do not form a clot in the abomasum and so slow down digestion. There are, however, people who have got away with both of these practices.

'MILKLESS' POWDERS

Powders based on whole milk or skimmed milk are expensive, although as I have said already, giving a calf a good start seems to me to be money well spent. In the past, attempts to substitute vegetable protein and carbohydrate for the casein and lactose in skim-milk have been unsuccessful unless these were introduced gradually and relatively late in development, as in the milk and gruel system (Table 7). The main obstacle to the development of a successful milkless powder for rearing calves before weaning has been the problem of finding a successful substitute for milk proteins. There is no particular need to find a substitute for lactose as it is a plentiful by-product of cheese-making. As indicated above, the albumin and globulin in whey can provide an acceptable substitute for casein, particularly in strong acid feeds, but vegetable proteins have tended to fail on two counts. The first, obvious reason is that vegetable proteins are less well digested by the young calf than milk proteins and so growth rates and food conversion efficiency are both inferior to those of calves on whole milk. Secondly, it has been shown in recent years that some of the most popular protein feeds for livestock, e.g. soya flour, left after conventional industrial procedures for extracting oil from the soyabean, are highly antigenic when fed to young calves (Killshaw & Sissons, 1979). These antigens may set up severe food allergies involving widespread damage to the absorptive surfaces of the epithelium of the small intestine. Interference with the mechanisms of secretion and absorption across the wall of the small intestine causes, at the least, a reduction in digestibility of food; at its worst, it leads to catastrophic diarrhoea.

Early studies of the digestion of vegetable proteins by calves did not distinguish between poor digestion due to enzymic incompetence, and poor digestion due to epithelial damage. As our understanding of these things progresses, it appears that if vegetable proteins are treated in such a way as to destroy the food antigens, their digestibility may be acceptably high, though not as good as milk protein. At present the costs of treating vegetable proteins to destroy food antigens tend to be high and so raise the cost of milkless powders close to that of those based on skim-milk. If the cost of treating conventional vegetable proteins like soya can be reduced or if a suitably non-antigenic vegetable protein source can be found, then 'milkless' powders (which really means, powders containing little or no milk protein) could be sold at least £100/tonne cheaper than conventional powders. Again, these powders would be best sold as strong acid powders for reasons explained in more detail later.

GRUEL

Gruel is best considered not as a low-cost alternative to milk powder, but as a starter ration fed as a drink rather than in solid form. As Table 7 shows, it should not be introduced until the third week of life and thereafter it is fed up to about 7–12 weeks of age as a partial substitute for the starter ration that an early-weaned calf would be consuming at this time. Gruel usually consists of 240 g/kg protein, mostly of vegetable origin, and about 55 g/kg fat. Recipes for gruel often include a barrage of different ingredients. This practice obviously evolved empirically, but with hindsight, this may have reduced the risk of clinical food allergies arising from excessive consumption of any one antigen. A few recipes for simpler home-made gruels are given in Table 8. For more complete details of these, see Roy (1980).

Linseed cake meal is the main protein source in each of these rations, presumably because it is so palatable. The high content of linseed oil in Ration 3 is not strictly necessary but it undoubtedly

Table 8 Composition of some typical home-made gruels (Parts by weight)*

	Ration		
	1	2	3
Ground oats	1	2.5	4
Fine wheat middlings	1	2.5	–
Maize meal	–	2.5	–
Linseed cake meal	3	1.5	4
Blood meal	1	1	–
Dried whey powder	2.5	–	–
Dried skim-milk powder	2	–	–
Ground linseed	–	–	2
Nutritive value			
Crude protein (g/kg)	280	250	260
Oil (g/kg)	35	40	120

*These rations should be supplemented by 20 g/kg of sterilised steamed bone flour

does help to put a 'bloom' on the coat of calves and they would look very nice to a prospective buyer at about twelve weeks of age.

FERMENTED COLOSTRUM

The modern dairy cow produces in the first three days of her lactation enough colostrum and first milk to provide her calf with two bucket-feeds per day to weaning at about five weeks. Moreover, the popularity of the 'beesting pudding' is not what it was. Consequently, a lot of colostrum gets thrown away. Dairy farmers in the U.S.A. therefore adopted the practice of storing colostrum in containers around the farm, allowing it to sour and then feeding the resulting smelly mess to calves twice daily. Some calves are extremely reluctant to approach the stuff but a rather surprising proportion consume it readily. Research in the U.S.A. and at Bridgets Experimental Husbandry Farm has attempted to refine the technique somewhat by using preservatives to control the fermentation, but these refinements have not really affected palatability or calf performance which, at best, is comparable to that of calves given equivalent amounts of whole milk.

The system is probably best suited to the dairy herd in which cows calve in the late autumn and winter. Colostrum taken from cows at this time and stored in plastic buckets at about outside air temperature will sour but remain reasonably acceptable for about three weeks. It would be unrealistic of me not to acknowledge that its keeping qualities may sometimes be improved by the addition of antibiotic-contaminated milk from mastitic cows. If they will drink it, calves can be reared to weaning with fermented colostrum as their only liquid feed. Another successful method has been to feed it alternately with conventional milk powder, i.e. a colostrum feed in the morning when calves are likely to be hungry and a conventional powder feed in the evening.

CALF STARTER RATIONS

The declaration of contents on the bag containing most commercial calf starter rations reads, Oil 3.5%, Protein 18%, Fibre 6.5%. This provides Metabolisable Energy at about 12 MJ/kg as fed (M/D = 13.5). The starter may be provided in the form of small pellets or as a coarse mixture of ingredients such as rolled oats, flaked maize and linseed cake. Coarse mixes appear to be more attractive

Table 9 Suggested compositions for calf starter rations (kg/tonne)

	Ration		
	1	2	3
Rolled oats/barley	275	300	200
Flaked maize	540	400	300
Whole fish meal	15	100	–
Soya or groundnut meal	35	50	–
Linseed cake	–	–	200
Dried skim-milk	25	–	–
Molasses (or molassine meal)	100	140	75
Mineralised balancer pellets	–	–	225
Mineral/vitamin supplement*	10	10	–

*A proprietary mineral mix should be added to rations 1 and 2 at the rate directed by the manufacturers (usually 10 kg/tonne)

to calves than pellets and so are particularly suitable for calves given *ad lib.* access to milk replacer from teats, since they are in particular need of a counter-attraction.

Three specimen compositions for home-mixed calf starter rations are shown in Table 9. Rations 1 and 2 are completely home-mixed except for the proprietary mineral/vitamin mix which is essential to balance the diet for (especially) sodium, calcium, phosphorus and vitamins A, D and E. Ration 3 is a simpler mix based on three readily available ingredients supplemented at a rate of 225 g/kg by a commercial balancer containing not only minerals and vitamins but also high quality protein from white fish meal and skim-milk. Oats are probably more palatable than barley to the young calf and also safer, as they are fermented more slowly. However, barley is the only one of these ingredients that is likely to be grown on cattle farms. Rolled home-grown barley can be substituted for oats provided the calves have plentiful access to long roughage and fresh water (see below) but it may carry a greater risk of causing ruminal acidosis or bloat.

STARTER ROUGHAGE

Conventional wisdom has it that the best form of roughage for young calves is soft meadow hay. In fact the best form of roughage is, of course, fresh spring grass but dairy calves are unlikely to see this as youngsters. The attractions of soft meadow hay are that it encourages proper development of the rumen and that it encourages the processes of rumination and particularly salivation. The calf that ruminates and salivates profusely naturally produces large quantities of water and sodium bicarbonate which dilute the rumen contents and prevent them from becoming too acid as a consequence of rapid fermentation of the starter concentrates. Sodium bicarbonate has been added to starter rations (at 35 kg/tonne) to prevent excess rumen acidity but the evidence to date suggests that the natural effect of saliva, which both dilutes and buffers the rumen contents, is more effective.

Although barley straw has little nutritional value for the young calf, it too stimulates rumination and salivation and so aids the orderly fermentation of concentrates. Hay or barley straw is perhaps best fed in the long form to encourage initial chewing and subsequent rumination. Hay or straw chopped to a length no less than 5 cm still stimulates rumination but if hay or silage is chopped very short, the calf has little need to ruminate and the risk of ruminal acidosis and bloat is increased.

Feeding systems compared

PERFORMANCE AND COST

Table 10 compares the performance and cost of rearing calves on five different systems based on a study made at Liscombe Experimental Husbandry Farm in 1980.

All calves were purchased at approximately one week of age and weaned at about 60 kg liveweight. This technique was employed to standardise conditions for the different rearing systems since it was, of course, impossible to monitor individual intakes of starter rations in the group-reared calves. Time to weaning was about 35 days except for the once-daily bucket-feeding cold milk system. The unrestricted teat-feeding systems gave the best 12-week weights. Calves drank the same amount of cold, mildly acid milk as they did warm milk from an automatic dispenser. Bucket-feeding twice-daily warm milk was the most efficient system in terms of cost per kg gain. Bucket-feeding once-daily cold milk significantly reduced performance, and in the end worked out the most expensive method. All the other rearing systems could really be deemed satisfactory given good strong calves, the excellent quality of the stockmanship at Liscombe and a certain amount of good luck, particularly in

Table 10 Comparison of calf feeding systems
(Figures from Liscombe Experimental Husbandry Farm (1981)

Rearing method	Days to wean	12-week wt (kg)	Milk powder (kg)	Early wean pellets (kg)	Hay (kg)	Food costs 0–12 weeks (£)	Cost/kg gain (p)
Bucket-feeding twice-daily – warm	34	100	14	118	12	30.16	52
Bucket-feeding once-daily – warm	34	95	15	108	12	29.13	55
Bucket-feeding once-daily – cold	45	90	15	112	12	29.81	62
Automatic dispenser warm milk	34	106	30	100	12	37.82	60
Unrestricted mild acid cold milk	35	105	29	105	12	38.00	60
High fat milk powder			£670/tonne				
Early weaner concentrates			£170/tonne				
Hay			£60/tonne				

All calves were purchased at one week old weighing on average 42 kg and weaned at approximately 60 kg liveweight regard to the ever-present risk of buying in disease. The control of disease in the bought-in calf will be discussed in detail in Chapter 6. At this stage it is necessary to speculate in rather general terms on the implications of these different feeding systems on the incidence and severity of enteric diseases in young calves.

DISEASE CONTROL

When calves are taken from their farms of origin, mixed and transported through markets to specialist rearing units, it is inevitable that some will become infected with one or more pathogenic microbes, the commonest of the severe pathogens being *Salmonella typhimurium*. One of the main reasons advocated for rearing calves in individual pens with solid sides stretching forward of the front of the pens (Fig. 5(a)) has been to stop calves licking one another and spreading *S. typhimurium* or other enterobacteria by direct contagion. If calves are reared in groups and drink from a communal teat, spread of infection by direct contagion becomes inevitable. If each calf that was infected with *S. typhimurium* or another pathogenic enterobacterium showed clinical signs of septicaemia – high temperature, dullness, scouring, etc. (see Chapter 6) – then one would expect to see a far higher incidence of scours and septicaemia in calves reared in groups than those reared in isolation. In fact this does not seem to be the case. Table 11 compares bucket-reared calves in individual pens with calves reared in groups and fed either warm milk from an automatic dispenser or acidified cold milk, in terms of death rates and numbers of calves treated for diarrhoea to 12 weeks of age. The data comes from two recent (1982) surveys by British Denkavit Ltd. and by the State Veterinary Service in association with the University of Bristol.

The survey by the State Veterinary Service and the University of Bristol primarily involved a study of the development of behaviour (Chapter 7) and was based on too few animals to reveal statistically significant differences in death and disease between groups, but the figures do suggest that the incidence of death and enteric disease in group-rearing systems was acceptably low. British Denkavit have monitored the performance of a far greater number of calves reared in groups and teat-fed from automatic dispensers. They do not present any comparable data for individually fed calves but the death rate and incidence of diarrhoea in bought-in calves especially are certainly quite as good as a calf-rearer would expect to achieve with calves in individual pens.

Table 11 Comparison of calf feeding systems in terms of death rates and numbers of animals treated for diarrhoea to 12 weeks of age

Survey	System	Number of calves	Deaths		Treatment for diarrhoea	
			No.	%	No.	%
State Veterinary Service University of Bristol	Individually penned, bucket-fed					
	home-reared	108	nil	nil	4	3.7
	bought-in	251	2	0.8	6	2.4
	Home-reared, in groups					
	automatic dispenser	128	1	0.8	9	7.0
	cold acid milk	55	1	1.8	nil	nil
British Denkavit	Automatic dispenser, groups					
	home-reared	9403	187	2.0	737	5.2
	bought-in	4781	95	2.0		

The question arises therefore, 'Why is the incidence of clinical septicaemia apparently unrelated to the opportunity for contagion?'. There is no clear-cut answer to this but one can make an intelligent guess based on an understanding of how enteropathogens get into the small intestine in sufficient numbers to do damage. Consider a young calf exposed for the first time to a particular species and strain of pathogenic micro-organism, say *S.typhimurium* or an enteropathogenic strain of *Escherichia coli (E.coli)*. If this organism is drunk it will probably enter the abomasum. Whether it survives passage through the abomasum into the duodenum and small intestine depends on the degree of acidity in the abomasum at the time. As a general rule it can be said that if abomasal pH is below about 4.7, most bugs entering the abomasum will die there. When the abomasum is empty or nearly empty of milk, the contents are extremely acidic (pH 2). What happens after the calf gets a meal depends both on how much it drinks and the nature of the feed. Figure 7 illustrates what happens to abomasal pH in a calf getting a simple bucket feed of about 2.5 ℓ of whole milk or milk replacer. When the calf drinks whole milk at blood heat, abomasal pH initially rises rapidly to neutrality (pH 7) and then declines in response to secretion of hydrochloric acid (HCl) by the chief cells of the abomasal epithelium. About 3 h after the meal, abomasal pH has fallen once again below 4.2 and abomasal killing of most bacteria can be guaranteed.

Suppose, instead, the calf is given the same quantity of a conventional 'sweet' milk replacer fed at air temperature. In this case, the decline in abomasal pH is slower. The stimulus to HCl release is better when whole milk is fed than reconstituted milk powders and better when milk is fed warm rather than cold. In the example shown in Fig. 7, abomasal pH stays above 4.2 for 4 h following the meal of cold milk replacer.

When the milk replacer is acidified in advance, abomasal pH will not, of course, rise any higher than that of the feed. Natural secretion of HCl tends to be lower following ingestion of acid feeds than whole milk, but even so the period during which most microbes can survive in the abomasum is markedly reduced by milk acid feeds and eliminated altogether when strong acid feeds are given. Clinical impressions are that the risk of scouring is higher when milk replacers are fed, rather than whole milk, and higher when milk is fed cold rather than warm. It is reasonable to explain this in terms of the length of time that the small intestine is at risk

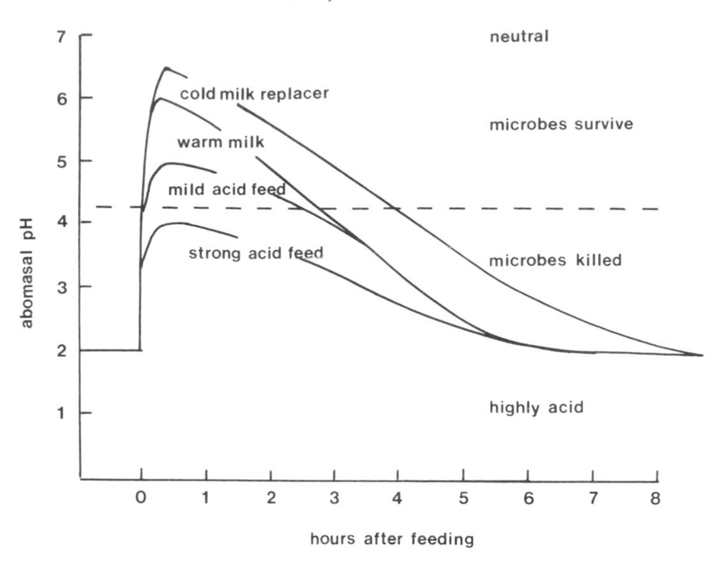

Fig. 7. Effects of abomasal acidity (pH) of a single feed of 'sweet' milk replacer (cold or warm) 'mild' acid and 'strong' acid replacers. For further explanation see text.

from attack by pathogenic micro-organisms that have survived passage through the rumen. By the same token, acid feeds must be deemed 'safe' because they ensure a good abomasal kill.

If conditions inside the abomasum are always such that microbes are likely to get killed *en passant*, then the need to form an abomasal clot and so slow down the passage of food largely disappears. This is why it has been possible to include in strong acid milk replacers large quantities of whey proteins and vegetable proteins which, if incorporated into conventional 'sweet' milk powders, might have seriously increased the risk of bacterial enteritis.

Most of the bacteria entering the abomasum of the young calf actually come from the rumen. The vast majority of the micro-organisms passing out of the mature rumen are completely harmless and, following digestion in the abomasum and duodenum, serve as the major source of dietary protein for the adult cow. This normal rumen population of commensal micro-organisms competes moreover with potentially pathogenic enterobacteria so that, while the concentration of *E. coli*, for example, may be very high in the rumen of the unweaned calf (or the veal calf), these organisms are usually entirely absent from the rumen of the adult animal. In the unweaned calf, therefore, the majority of potentially pathogenic micro-organisms entering the abomasum have not come direct from the mouth but have probably arisen from a colony, previously ingested but now thriving in the medium of the immature rumen. Maturation of the normal anaerobic, cellulose-fermenting rumen population kills off organisms like *E.coli* and this is almost certainly why problems of bacterial enteritis in calves tend to disappear with weaning.

Another way for the unweaned calf to keep abomasal pH low enough to sterilise the digesta passing through is to drink little and often. This is what *most* calves do when given free access to milk from their mother or warm sweet milk replacer from an automatic dispenser. This again would explain why the majority of machine-fed calves do not succumb to enteric disease even when pathogenic strains of *Salmonella* or *E.coli* are known to be in the unit. Bucket-fed calves reared in groups may be more vulnerable because they drink more milk at one go.

The other way that enterobacteria can attack the delicate tissues of the small intestine is from the rear. It is often possible to isolate such bacteria from rectal swabs taken from clinically healthy animals. In these circumstances we must assume that these organisms are existing in small numbers in the large intestine and last third of the small intestine, their numbers being kept low by the fact that the wall of the small intestine is undamaged and that very little undigested food is getting down to them. In short, their population numbers are being constrained by starvation. If, however, the calf overloads with food, or gets an attack of indigestion for any reason, then undigested food will enter the terminal ileum, allow these bacteria to multiply and advance up the small intestine, finding more and more food and multiplying as they go. Again, if the epithelium of the small intestine is damaged, even slightly, by food antigens or by viruses, then opportunist pathogens like *E.coli* can exploit this damage by growing on the damaged tissue and liberating their own toxins (or poisons). Seen in this light, it is not difficult to appreciate how bacterial enteritis in unweaned calves can result from (a) overfeeding, (b) poor quality food, or (c) a change of diet.

Manufacturers of milk powders for calves are aware of these problems, and so select their ingredients and formulate their diets with great care. It is not often, therefore, that one can simply blame 'the feed' for causing enteritis in calves. However, although manufacturers do their best to maintain consistency between successive batches of feed, not only in terms of the quantity of fat and protein contained in the bag but also the quality of the protein and, for example, the particle size of the fats in emulsion, they cannot hope to subject each feed to the same specific and scrupulous examination as it gets from the enzymes inside a calf's gut. It is thus very common for calves to show some signs of indigestion after they are given a different milk powder or even a new batch of the same brand of milk powder. It is therefore good husbandry to rear calves right through to weaning on the same batch.

The occasional 'compulsive drinker' is always likely to be a problem in any conventional teat-feeding system. Acidifying the milk powder does not appear to reduce overconsumption in such calves, who may drink 20 ℓ/d or more for reasons that clearly have little if anything to do with hunger or thirst in the strictest sense of the word. Such calves are not only very prone to diarrhoea in the first instance but also prone to relapse after treatment since this should always incorporate feed restriction (Chapter 6). The farmer who operates a free-choice teat-feeding system is bound to get the odd 'compulsive drinker.' If such an animal does not settle down within a few days it pays to take it out of the group and rear it on a bucket. In a large specialist unit rearing calves in groups, the compulsive drinker becomes one of the best justifications for the computerised calf feeder that takes away the teat when the calf has had enough.

Further reading

Killshaw, P.J. & Sissons, J.W. (1979) 'Gastro-intestinal allergy to soyabean proteins in pre-ruminant calves. Antibody production and digestive disturbances in calves fed heated soyabean flour'. *Research in Veterinary Science*, **27**: pp. 361–365.

Roy, J.H.B. (1980) *The Calf.* 4th ed. Butterworths, London. pp. 276–280.

4 Environmental needs

The essential environmental needs of a young calf can be summed up in three words – comfort, space and health. This simple definition can be expanded for the purposes of proper analysis as follows:

Comfort

(a) *Thermal comfort.* The environment must not be so hot or so cold as significantly to affect production or cause distress.
(b) *Physical comfort.* The space available to the calf and the floors and surfaces with which it makes contact should be such as to minimise the risk of acute injury or chronic discomfort.

Space

The space available to each calf should *at least* give it sufficient room to 'stand up, lie down, turn round, stretch its limbs and groom itself' (Brambell, 1965). It should also provide calves with a reasonable degree of social intercourse.

Health

The environment should be designed in such a way as to minimise disease. This may be achieved by (a) hygiene, i.e. by reducing the spread of infection by direct contact (contagion) and by airborne transmission, and (b) avoiding stresses liable to decrease resistance to infection.

These needs seem modest enough. However, there are, in fact, very few calf-rearing units that meet them all. Indeed, they are to some extent incompatible. The calf reared in an individual pen with solid sides (Fig. 5(a)) is denied social intercourse in the interests of hygiene. Veal calves reared in groups for 'welfare' reasons (Chapter 8) are undoubtedly exposed to a greater risk of infection than those reared in crates. Trying to keep calves warm may increase the risk of pneumonia. The 'best' environment has to be a compromise between the various needs of the calf and other matters of importance, such as ease of operation and costs of the building.

However, in housing as in feeding, it is first of all essential to understand how the calf works.

Thermal comfort

TEMPERATURE REGULATION

The calf, like all mammals, is a homeotherm; that is to say it maintains a constant deep body temperature, in this case at about 38.6° or 101.6°F. To achieve this it must balance the amount of heat it produces in metabolism (H_p) against that which it loses to the environment (H_l). The heat balance equation is described by equation 4.1. For a more complete description of heat exchanges in cattle, see Monteith and Mount (1974).

$$H_p + H_s = H_l = H_n + H_e \qquad (4.1)$$

Where H_s is the amount of heat stored in the body per unit time (say 1 d)

H_n is the amount of heat lost by 'sensible' means, namely convection, conduction and radiation. The subscript 'n' is conventionally used to describe sensible heat loss because it obeys Newtonian laws, i.e. it is proportional to the temperature gradient from the calf to the environment.

H_e is the amount of heat lost by evaporation of water from the skin and respiratory tract.

The units for heat transfer may be MJ/d which is useful in the context of calculating daily requirements for Metabolisable Energy. In the context of building design it is more useful to describe heat transfer in watts (J.sec^{-1}).

Sensible heat loss from the site of metabolic heat production in the organs and tissues of the body (the body core) passes through two layers of insulation provided (i) by the tissues of the body, and (ii) by the hair coat (Fig. 8(a)).

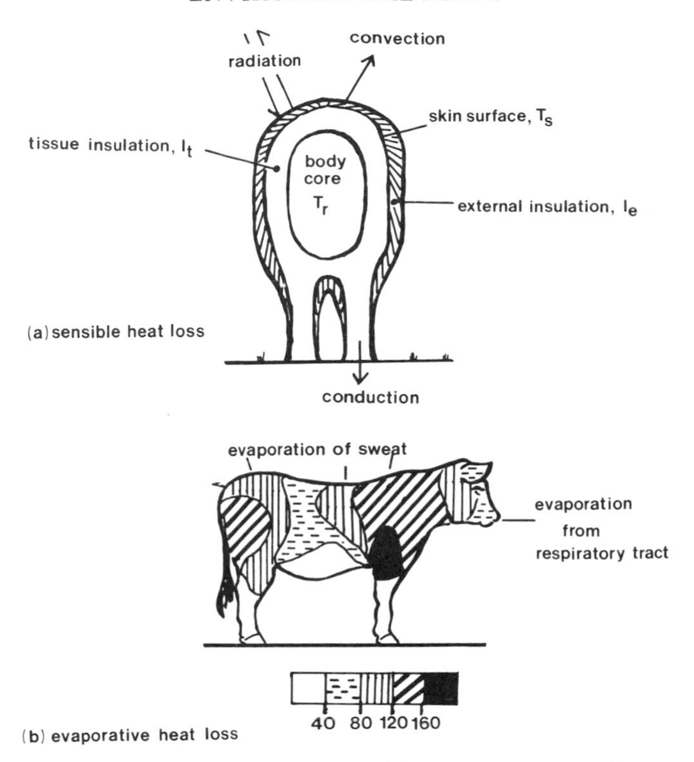

Fig. 8. Pathways of heat loss from calves. (a) Sensible heat loss by convection, conduction and radiation from the body core (T_r) through two layers of insulation, tissue insulation (I_t) and external insulation (I_e). (b) Evaporative heat loss from the skin and respiratory tract. The shading indicates regional differences in maximum sweating rate from below 40 to above 160% of the average.

Tissue insulation describes the resistance to heat loss provided by the superficial tissues of the body shell. The calf can regulate heat loss by convection through the body shell by constricting or dilating blood vessels under the skin. When the calf feels warm the superficial blood vessels dilate to carry body heat to the surface. When it feels cold the superficial blood vessels constrict and heat loss through the body shell is reduced. If the temperature is cold, but not freezing, heat passes through the body shell principally by conduction. In this case, tissue insulation (I_t) is determined by the thickness of the skin and the subcutaneous fat, and can be written

$$I_t = (T_r - \overline{T}_s) \cdot A/H_n \tag{4.2}$$

where T_r is rectal temperature (°C), $\overline{T}_s$ is the average temperature of the skin surface and A is body surface area (m^2). In practice, it is difficult to measure A, so it is usually predicted from body weight (W, kg).

$$A(m^2) = 0.09W(kg)^{0.67} \qquad (4.3)$$

If air temperature is below 0°C, the calf cannot fully constrict the blood vessels of the body shell or exposed extremities like the ears and feet or they would freeze, since they produce little heat themselves and rely on their blood supply to keep them warm.

External insulation (I_e) describes the resistance to sensible heat loss from the skin ($\bar{T}_s$) to the air (T_a) by the coat of hair and (mainly) the layer of warm air trapped in it and on it. Thus,

$$I_e = (\bar{T}_s - T_a) \cdot A/H_n \qquad (4.4)$$

In Fig. 8(a) where the animal is standing up, H_n passes from skin to air by convection and radiation. When the animal lies down, much heat is lost by conduction to the ground and the insulation of the surface on which it lies assumes major importance.

The major factor determining I_e is, of course, the depth and density of the hair coat. There would appear to be clear differences between breeds of cattle in this. On the whole, beef breeds have thicker coats than dairy breeds. Highland cattle are an extreme example but Hereford, Galloway and Angus cattle out on the hill very obviously have thicker coats than, say, Friesian cows in a cubicle house. Most of this difference is not, however, genetic. Cattle have considerable ability to adjust coat depth according to how hot or cold they sense the environment to be. The way in which they do this is rather subtle. Hair growth is predominantly under the control of light. Thus, in the autumn, hair growth rate increases to prepare for the winter. Hair shedding is, however, determined by temperature, or more precisely by the sensation of heat or cold as perceived by each individual. Hair shedding rate determines how long each hair grows before it falls out and is therefore the factor which ultimately governs coat depth. As we shall see later, one of the main factors that determines how warm or cold an individual cattle beast feels is the amount of heat it produces as an inevitable consequence of metabolism and *this* is primarily determined by the amount of food it gets to eat. The main reason why Friesian dairy cows have short, sleek coats summer and winter is that they get so much to eat that they don't feel cold. The animal that eats little, perhaps because it has been sick or simply because food is not available, develops a long thick winter coat and keeps it into the spring. It is often said that the 'poor doer' is the last to lose its winter coat. Veal calves can get very hot because they take in so much food energy and they tend to shed their birth coat at 8–10 weeks of age.

Wind, rain and snow all increase H_n at a given temperature gradient from skin to air ($T_s - T_a$) by reducing the insulation of the coat. Incoming radiation from the sun has the effect of reducing net H_n. The mathematics of heat transfer of cattle out of doors are too complex to be described here in detail (but see Webster, 1971). It is, however, important to the design of calf houses to know the effects on H_n of (a) draughts (i.e. increased air movement on the calf surface), and (b) different floor and bedding materials. These will be considered later.

To summarise: sensible heat loss from an animal is primarily determined by the temperature gradient from the body core to the air. The rate at which H_n increases with increasing ($T_r - T_a$) is determined by tissue and external insulation. Figure 9, which plots H_p, H_n and H_e in a 6-week-old calf as a function of T_a, shows that H_n increases rapidly as T_a falls from 35 to 20°C, because in this range the calf is warm, the superficial blood vessels are dilated and I_t is low. Below about 20°C the rate of increase drops as the calf constricts the superficial vessels (increasing I_t) then erects the hair coat (slightly increasing I_e).

EVAPORATIVE HEAT LOSS

Cattle have two very effective methods for regulating heat loss by evaporation (H_e). First of all, they can sweat copiously and for long periods. Regional differences in sweat rate from the skin surface are shown in Fig. 8(b). It is, of course, not the production of sweat but the evaporation of sweat that cools the animal, so that if (a) the humidity in the building is very high, or (b) the sweat gets trapped in a thick winter coat, then the efficiency of this mechanism is impaired.

The second mechanism for regulating H_e involves what is usually referred to as thermal panting. It is well known that cattle increase respiration rate when they get hot. The purpose of this is to ventilate the large vascular surface area over the turbinate or 'scroll' bones in the nose which act as highly efficient heat exchangers, particularly for the blood supplying the brain. In a cool or cold environment a calf or cow at rest will breathe deeply at about 18–20 respirations per minute. As air temperature increases so the calf will breathe faster but less deeply increasing air flow over the heat exchangers in the nose but keeping air flow to the lungs constant. This enables it to lose more heat without affecting the exchange of oxygen and carbon dioxide. The upper limit of respiration rate during thermal panting is about 180/min.

The respiration rate of calves provides a most useful simple indicator of whether they are warm or cold. On warm days it is usually easiest to count movements of the flank in calves that are lying down. On days that seem to us to be cold it is often possible to count the puffs of steam coming from their nostrils. If the respiration rate of calves is 20/min or below, they are at least cool, and could be very cold. If, however, the respiration rate of healthy calves is over 25, they are not cold whatever the environment may feel like to us. It is common to record respiration rates of about 50–60/min in strong, well-grown calves in a follow-on yard in mid-winter. Such calves are very comfortably warm. Figure 9 shows H_e falling with decreasing T_a to a minimum at about 10°C. This minimum value for H_e, about 17 watts/m^2 surface area, represents the inevitable loss of water vapour from the skin and respiratory tract when active sweating has been switched off and respiration rate reduced to a minimum.

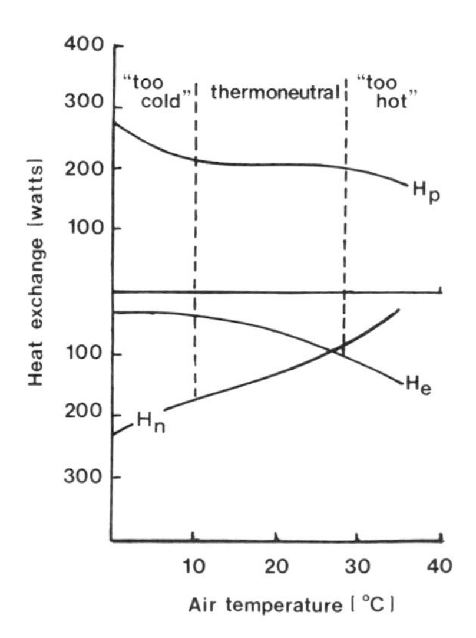

Fig. 9. Heat exchanges of a 90 kg calf at air temperatures from 0 to 40°C. Heat production (H_p) equals the sum of sensible (H_n) and evaporative heat loss (H_e). For further explanation see text.

METABOLIC HEAT PRODUCTION

In most practical circumstances all the heat produced by a calf arises as an inevitable consequence of the work it has to do in metabolism. Its tolerance of cold and sensitivity to heat are governed by the amount of heat produced per unit of body size, or more precisely surface area (equation 4.3), and this is primarily determined by the amount of food energy it eats. Other things being equal, the calf that gets the most to eat is the most resistant to cold stress. The calf that has been deprived of food by reason of disease, or transport, has the lowest heat production and so is the most susceptible to chilling.

In the example illustrated in Fig. 9, H_p is unaffected by T_a within the range 10–30°C. This is the *thermoneutral zone* wherein H_p is unaffected by variations in T_a and homothermy is maintained by physiological regulation of H_n and especially H_e. The thermoneutral zone thus defined equates with the zone of reasonable thermal comfort and also, more or less, with the zone of optimal food conversion since no food energy is 'wasted' in keeping warm.

The upper limit of the zone of thermal neutrality is reached when the calf finds that it cannot comfortably dissipate by evaporation the amount of heat that it is producing in metabolism and so takes the next step essential to maintaining homothermy which is to reduce H_p by reducing food intake. Reduced heat production through reduced food intake is one of the major constraints on cattle production in the tropics (Hafez, 1968). The possibility of young calves in the U.K. suffering heat stress is extremely remote. It is only likely to occur if they are exposed for prolonged periods to direct sunlight.

The lower limit of the thermoneutral zone, the *lower critical temperature*, signifies the air temperature below which the animal must elevate H_p in order to maintain homothermy. This is important for two reasons. As T_a falls below the lower critical temperature, so food conversion efficiency decreases because more food energy goes towards keeping the calf warm. Furthermore, the intensity of cold stress, measured by the elevation that it induces in H_p, is likely to affect the ability of the calf to resist infection. The lower critical temperature (T_{lc}) can be calculated from expressions described above.

$$T_{lc} = (T_r + H_{e, min} \cdot I_e) - H_{min} (I_t + I_e) \quad (4.5)$$

where H_{min} is thermoneutral heat production (watts/m^2)

$H_{e, min}$ is minimal evaporative heat loss in the cold (watts/m^2) (Fig. 9)

The units of I_t and I_e are °C.m^2/watts.

COLD TOLERANCE

Table 12 presents values for the cold tolerance of calves from birth to six months of age. The figures relate to a single calf standing up in dry conditions, (i) with negligible air movement, the sort of conditions one would hope to achieve in a well-designed but unheated calf building, and (ii) at an air movement of 2m/sec (about 4.5 miles per h).

Table 12 shows that there is not much difference in cold tolerance between beef-type (e.g. Hereford) and dairy-type (e.g. Friesian) calves in early life when they are feeding properly, the slightly greater thermal insulation (I_t + I_e) of the beef calves being offset by the slightly greater thermoneutral heat production (H_{min}) of the dairy calves. The greater thermal insulation of the beef animal only really confers any practical advantage when it is on open range and getting relatively little to eat (i.e. the beef cow). This is however outside the scope of this book (see Webster, 1971).

The lower critical temperature of the newborn calf, after it has dried off, but before it has really begun to digest and metabolise food, is a little higher than 10°C (50°F). This is not to say that it is severely stressed by cold if air temperature is below 10°C, merely that it starts to feel cold at about this point, a much less alarming concept. The calf that has recently arrived from market and is deprived of food while it 'settles down' is as sensitive to cold as is the newborn animal and for the same reason: H_{min} is low because no food is getting absorbed and metabolised.

By the time it reaches weaning age (5 weeks), the normal healthy calf, eating well, has acquired a very respectable degree of cold tolerance and is quite unaffected by air temperatures above freezing point provided that it is dry and not exposed to draughts. By 6 months of age its cold tolerance has become even more impressive, again provided that it is getting plenty to eat. It cannot be stressed too strongly that a starved calf is a cold calf. Since both stresses have the same effect, which is to exhaust the animal's energy reserves, they are at their worst when they occur together. Perhaps the most startling demonstration of the effect of food intake on cold tolerance is provided by the veal calf in the later stages of development (say,

Table 12 Factors determining the cold tolerance of calves

	Thermoneutral heat production (W/m^2)	Thermal insulation ($^\circ C.m^2/W$)		Lower critical temperature ($^\circ C$) at air speeds (m/s)	
		I_t	I_e*	0.2	2.0
Newborn calf					
Beef	100	0.09	0.23	+11	+19
Dairy	100	0.08	0.23	+12	+20
Weaner calf (5 weeks)					
Beef	115	0.10	0.26	+ 2	+10
Dairy	125	0.09	0.24	+ 2	+11
Weaned calf (6 months)					
Beef	120	0.14	0.28	− 6	+ 5
Dairy	130	0.12	0.25	− 6	+ 5
Veal calf (12 weeks)	155	0.12	0.23	−11	+ 2

*In still air (0.2 m/s)

12 weeks of age). By this time it is taking in so much food energy that it is extremely tolerant of cold, and sensitive to heat, despite the fact that it will, by 12 weeks, have shed its birth coat and have the sleek coat of the adult dairy animal. Many veal rearers on the continent of Europe clip the hair along the back of their calves at about 8 weeks of age to help them to dissipate heat. In the first few (4–5) weeks of life the veal calf is no more or less sensitive to cold than a conventionally reared animal.

Table 12 also illustrates the extent to which T_{lc} is elevated by increasing air movement. Air speeds of 2 m/sec can occur in draughty regions of many conventional calf houses. If calves are reared in groups they may be able to get out of the draughts. If a calf is isolated in a small draughty pen there is nowhere for it to go. As a rule of thumb it is advisable to ensure that airspeed at calf height does not exceed 0.27 m/s (0.6 miles per hour).

FLOOR TYPE

When the calf lies down, two important changes take place in its pattern of heat loss. It now loses heat by direct conduction to the ground at a rate determined by the insulation of the material it is lying on. Secondly, it may be able to get out of draughts, particularly if it is able to nestle into deep straw. Table 13 illustrates the effect of different floor types on the lower critical temperature of calves, taking into account both of those effects. The example used for illustration is the newborn or recently bought-in calf which is most sensitive to cold.

Table 13 shows first that concrete makes an awful bed. Not only is it hard and abrasive, but it is extremely cold. Furthermore it is

Table 13 Effects of floor type on the lower critical temperature of a newborn calf

	Lower critical temperature (°C)
Calf standing	+ 11
Calf lying on dry concrete	+ 18
on wooden slats	+ 11
on damp straw	+ 11
in deep dry straw	less than 6

not possible to reduce heat loss through concrete by insulating *under* the screed; the conductivity is so great that heat simply escapes sideways (Bruce, 1979). A calf on wooden slats loses about the same amount of sensible heat standing up or lying down. However, when calves are housed on wooden slats in raised pens it is important to ensure that there is no possibility of draughts coming up through the slats. This is quite common, especially in solid-sided pens. A flat bed of damp, old straw is comparable to wooden slats, but free of the risk of draughts. A deep bed of fresh straw is, not surprisingly, the warmest because it reduces heat loss by conduction, cuts down draughts at calf height and traps a greater boundary layer of warm air around the animal. It has been calculated that, at least, the effect of deep, dry straw is to reduce T_{lc} for the newborn calf from 11 to about 6°C. However, physical calculations of the sort described in this chapter almost certainly underestimate the degree of thermal and physical comfort that a calf may obtain from a deep bed of dry straw in an unheated building, particularly if it is isolated in an individual pen. It has been claimed by observers of farm animal behaviour that calves, unlike piglets, are not a gregarious species, because they do not huddle together when they lie down. This argument misses the essential point which is that animals huddle principally to stay warm by sharing body heat. Piglets are usually observed at air temperatures below T_{lc}, calves usually at air temperatures above it. If piglets are kept warm, say by a heated floor, they spread out. On very cold days (below 0°C) calves reared in groups will huddle together.

Physical comfort

The design of the fixtures and fittings inside a calf house so as to minimise the risk of injury or chronic discomfort is largely a matter of common sense. The most important factor to consider in the context of physical comfort is, of course, floor design. The floor is the surface on which the animal stands, walks, lies down and passes excreta. It must, therefore, according to the needs of the moment, be unyielding, non-slippery and well-drained, or comfortably soft, warm and dry and easy to clean by machinery. No single material, from concrete to a grass meadow, meets all these specifications and it is probably fruitless to contemplate one.

The thermal properties of different floor types have been considered already. The next essential is that the floor should provide the calf with a secure footing. It should not be slippery, even when covered with urine and wet faeces. It should not yield under the weight of an animal weighing up to 100 kg. If slats are used, the gap between the slats should not exceed 2 cm. If that makes for some build-up of excreta, so be it. Calves don't produce a tremendous amount of excreta in the first few weeks of life, they don't stay in the rearing unit many weeks and it is easy (and essential) to clean out calf pens between batches with a hose or pressure washer. If the width between the slats is, say 4 cm, the calves can, in fact, stand up without their feet falling through, but they are very insecure when moving or standing up and lying down and they clearly find this distressing (see Chapter 7).

All in all, there really is no comparable substitute for straw to provide a warm, comfortable and secure bed for calves. Inadequate straw on a floor that doesn't drain can, however, be quite dreadful. Whenever possible, calf buildings should be designed so that the floor under the straw bed slopes away to an effective drain at a fall of 1 in 20.

Space requirement

As indicated earlier, the minimum space requirement for a calf was defined by the Brambell Commission (1965) as that which allows it, without difficulty, to stand up, lie down, stretch its limbs, turn round and groom itself. Four of these famous 'five freedoms' of Brambell are eminently reasonable. The freedom to turn round does not, however, seem to me to be terribly important as such. If, for example, calves are kept on suitable tethers for the first five weeks of their life before being weaned and turned out in groups into follow-on yards, this may not be too severe a constraint if the system (see Mitchell, 1978) helps to keep them free of disease. Tethering veal calves for life is another story (see Chapter 8).

Recommended minimum space allocations for calves are listed below. These provide reasonable freedom of movement, although the values for individual pens may prevent some calves from lying on their sides with their legs fully extended. How important this may be is discussed in Chapter 7.

Individual pens:	0–6 weeks, 1.5 x 0.9 m (1.35 m^2)
	0–8 weeks, 1.8 x 1.0 m (1.8 m^2)
Group pens:	0–8 weeks, 1.1 m^2
	0–12 weeks, 1.5 m^2

The floor area (m^2) recommended for calves in groups is less than for individual pens because calves can lie in any direction and so, in theory, no corners are wasted. It would however be a dangerous mistake to assume that rearing calves in groups is a good way to improve profit by increasing stocking density. As we shall see in the next section, the first constraint on calf numbers in a building is not floor space but air volume per calf. The values given above for group pens are best considered as minimal space requirements for the bedded area in a partially-bedded pen.

Health and Hygiene

All animals co-exist with a mass of potentially pathogenic micro-organisms. Indeed, infection may be said to be the natural state and disease merely the result of a loss of equilibrium between the animal host, the germ and the environment. The extent to which this generalisation holds true depends of course on the nature of the disease. In conditions like foot-and-mouth in cattle or smallpox in man, exposure to the germ is highly likely to produce clearly recognisable clinical signs of disease. This is one of the reasons why it has been possible to eradicate these two diseases (by rather different methods!). Many diseases of cattle have responded to control by the classic methods of eradication or vaccination and it is probably a reflection on the success of the veterinary profession that those of importance which remain (in the U.K. at least) are the non-specific 'environmental' diseases like the enteric and respiratory diseases of calves which appear to result from exposure to one or more of a wide range of micro-organisms coupled with some form of environmental insult. The effect of feeding systems on enteric disease has been considered already. It is now necessary to consider the effect of the environment in the building on disease in general and respiratory disease in particular.

The balance of infection is illustrated very simply in Fig. 10. The effect of the environment can be considered in slightly more technical terms (Table 14).

The environment may affect survival of the infectious micro-organism (the parasite) in the air or on the surfaces with which the calf comes into contact. It may affect the ability of the parasite to pass from one calf to another and to penetrate the respiratory tract. It may also affect the ability of the calf to clear the parasite from the surface of the respiratory tract (the first important point of

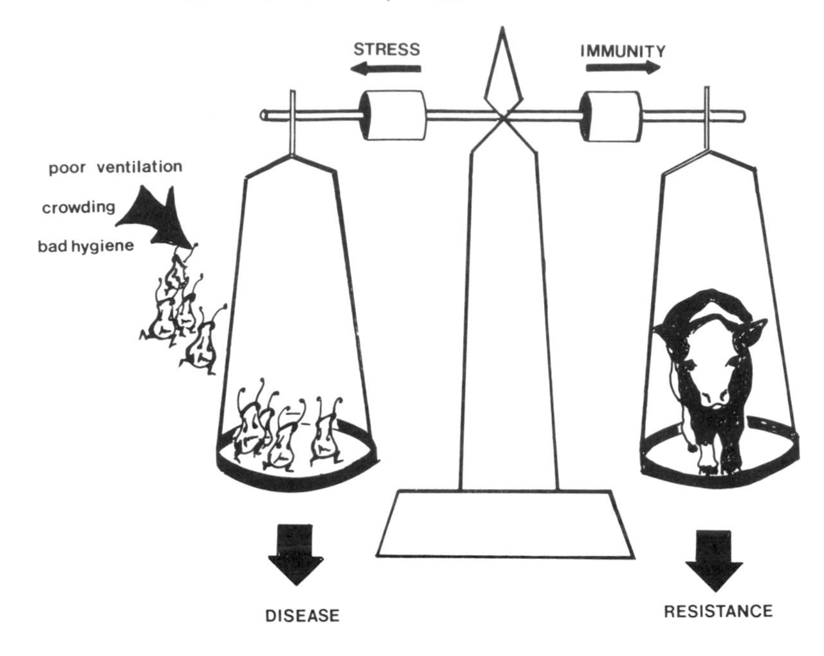

Fig. 10. 'The balance of infection'.

Table 14 Effect of the environment on infectious diseases of the respiratory tract of calves

Site of effect	Mechanism	Influential factors
Parasite	Survival in air	Air temperature, ventilation, relative humidity (RH), dust
	Survival on surfaces	Surface texture, disinfection
	Survival in bedding	Drainage, frequency of bedding change
Parasite/host interaction	Transmission	Stocking density, thermal environment
	Penetration	Air temperature, RH
Host	Local clearance	Air temperature, RH, air pollutants
	Systemic response	Stress (thermal, dietary, emotional), concurrent infection

contact between parasite and host). It may finally, by stressing the animal in some way, affect its ability to mount a defence against an acquired infection, thus increasing the severity of the infection to the point where the animal is obviously sick or, at worst, dies.

SURVIVAL AND SPREAD OF MICRO-ORGANISMS

Pathogenic micro-organisms may be transferred from one calf to another

(1) by direct contact (contagion); mouth to mouth, faeces to mouth, mouth to teat to mouth
(2) by short-range direct transmission of organisms in droplet form (coughing and sneezing)
(3) by airborne diffusion in aerosols (water droplets or dust)

Short of isolating calves completely from contact with other animals or with their own faeces, it is completely impossible to avoid some degree of contagion. Even when calves are isolated, if a virulent infection enters the building, it is, with the best will in the world, almost impossible to prevent indirect contagion via such things as the stockman's clothes or boots. The best that you can hope to do is to contain the infection within that building by thorough disinfection and by routinely changing into boots and protective clothing restricted to that building.

Short-range transmission of respiratory infection can occur when, for example, an infected calf sneezes or coughs and infected droplets pass directly onto the nose or mouth of another animal. This obviously depends on the proximity of calves to one another and the risk of short-range transmission of infection is obviously increased by increasing overall stocking density in the building so as to reduce the effective air space available to each calf. The risk is also increased if cold weather causes the calves to huddle or if some of the potentially available space is not used by the animals, for example, if it is wet or draughty.

The airborne transmission of respiratory infection occurs when calves breathe in organisms carried in aerosols of water vapour or of dust coming from such things as dry food and, especially, skin scales. One of the prime stated objectives of ventilating calf houses is to reduce the risk of pneumonia and good ventilation has become as important an item of the stockman's creed as good hygiene. Having already committed one heresy in the last chapter in questioning the absolute importance of hygiene in the control of enteric disease,

I must now compound it by questioning (but not rejecting) the importance of ventilation in the control of respiratory disease.

First of all, what does ventilation do? It removes from a building containing animals some of the end products of their existence; heat, water, carbon dioxide from respiration and other gases rising from the manure or slurry, and, of course, airborne organisms (bacteria, fungi and viruses). The effect of ventilation on heat and water loss is outside the scope of this book (see Sainsbury and Sainsbury 1979; Clark, 1981). Suffice it to say that if ventilation is enough to ensure, say, eight air changes per hour in a calf house, rather more than 90% of the heat and water produced by the calves and metabolism will be carried out in the ventilating air stream. The expression 'air changes per hour' is a useful way to describe the ability of the ventilation system to clear a building of, say, heat or germs. Eight air changes per hour means that the amount of air passing into and out of the building per hour is equal to eight times the volume of the building. In other words, the building is, in effect, being 'cleared' eight times each hour.

Recommendations for ventilating calf houses have generally attempted an uneasy compromise between the fact that, in a cold environment, ventilation on the one hand carries out useful heat produced by the calves themselves, or by artificial heaters installed within the building, but on the other hand must carry out lots of the germs released into the building. Mitchell's recent eminently sensible and useful *Calf Housing Handbook* (1978), recommends that minimal ventilation rate in winter should be 34 m^3/h per calf at a stocking rate of 6 m^3 air space per calf, which provides a clearance rate of $5\frac{2}{3}$ air changes per hour. This recommendation implies that clearance of micro-organisms is of prime importance since at this ventilation rate so much heat is convected out of the building that it is impossible within economic limits to keep air temperature in the building significantly above outside temperature. This reflects the wisdom of the moment, which is that it is much better to ensure that calves get plenty of fresh air than that they are kept warm. In view of what has already been said about their cold tolerance, this makes excellent sense. The heretic's question returns, however, in a new form: to what extent does 'fresh air' depend on ventilation?

To understand this we need a little more simple mathematics. The concentration of organisms in the air of any building depends on the balance between the rates of release of organisms into the air from the animals themselves and other sources such as the bedding, and rates of clearance of these organisms by, e.g. ventilation, sedimentation or by death in the air (Fig. 11).

$$(R_a + R_b) = C \times V(q_v + q_s + q_r + q_d) \qquad 4.6$$

Here R_a and R_b are rates of release of organisms (number/hour), V is the volume of the building (m^2) and C is concentration of organisms ($numbers/m^2$). Pathways of clearance (q) have the units (1/hour) and are as follows; q_v = clearance by ventilation, q_s = clearance by sedimentation, i.e. fall-out of organisms onto the building floor and walls, q_r describes organisms breathed out of the air into the respiratory tract of calves and q_d describes the rate at which micro-organisms are killed in the air by desiccation, ultra-violet light, etc.

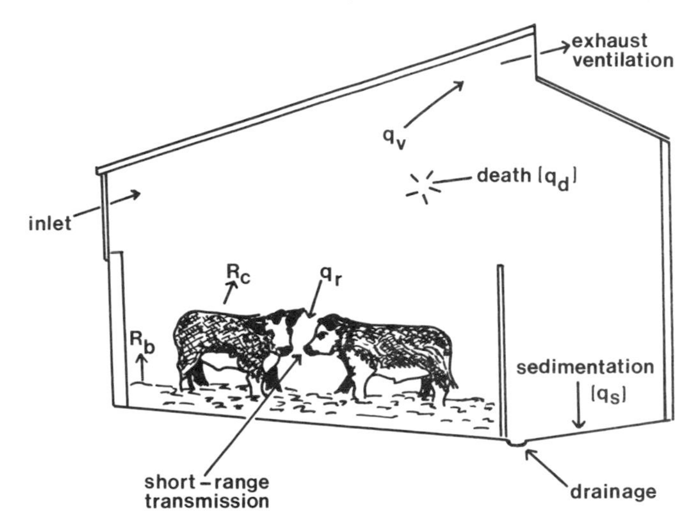

Fig. 11. Pathways for release, transmission and removal of infectious organisms within the air of an animal building (see Equation 4.6).

In a reasonably clean calf house the majority of organisms released into the air come from the calves themselves. Most of the organisms isolated from the air are not thought to be pathogenic. They are however an atmospheric pollutant and if their concentration in the air is very high, they may embarrass the calf's normal mechanisms for clearing harmful organisms from the respiratory tract. I shall return to this point later.

The most important point to realise at this stage is that since the calves themselves are the main source of airborne bacteria the most important factor determining the concentration of bacteria in the building is stocking density – the number of calves in a particular volume of building (*N*/*V*). Equation 4.6 can therefore be rearranged in a more useful form.

$$C = N/V \times R/(q_v + q_s + q_r + q_d) \qquad 4.7$$

Here, the concentration of airborne organisms (numbers/m^2) is expressed in terms of practical decisions as to management and housing design. *N*/*V* is stocking density; its reciprocal, *V*/*N*, is air space allowance (m^3 or ft^3) per calf. *R* here is rate of release of organisms per calf. The values for *q* describe the extent to which one can clear the air by the various routes illustrated in Fig. 11.

The concentration (*C*) of bacteria (or more precisely, of bacterial colony-forming particles, BCFP) in the air of a calf house can range from about 1000 to 500,000 BCFP per m^3. Out of doors *C* is about 100 per m^3; in a well-ventilated office it would be between 100 and 1500. The air inside a calf house tends therefore to be very polluted with BCFP (although less so than many intensive pig and poultry houses), the main sources being, as indicated already, the calves themselves.

Let us now examine the pathways of clearance. The first point to explain is that in normal circumstances the respiratory tract is a clearance mechanism. The turbinate bones in the nose and the complex plumbing of the lower respiratory tract provide a very effective filter, with the result that a healthy calf breathes out far less bacteria than it breathes in. From the calf's point of view, however, that can hardly be considered a good thing.

Clearance by sedimentation (q_s) depends on a variety of rather complex physical factors that need not concern us here since the sum of ($q_s + q_r$) is, in fact, quite small (about 3/hour) and not very important. Clearance by ventilation (q_v) depends, of course, on ventilation rate. At a ventilation rate of say 6 air changes/hour, particles released into the air would remain in the building for an average period of 10 minutes. The most important mechanism for clearing BCFP from the air of an animal building is q_d, death of the organisms while in the air of the building itself. The majority of BCFP are released into the air in droplets of water vapour or a particle of dust, especially skin scales. In most environments these aerosols rapidly dry out, the cell membranes of the bacteria rupture and the organisms die within a few seconds. Some organisms survive this first environmental shock but few survive in the air for more than a few minutes. One of the reasons that it is almost impossible to isolate pathogenic bacteria from the air of a calf house, even if it contains a number of animals with clinical signs of pneumonia involving, for example, *Pasteurella haemolytica*, is that such organisms die almost immediately on contact with the air. The organisms typically isolated from mixed air in a calf house are mostly those relatively harmless organisms which habitually live on the skin surface. For those organisms that survive initial exposure to air, q_d may vary between 30 and 100; for pathogens such as *P. haemolytica*, q_d may exceed 200. Mitchell (1978) recommends that to preserve fresh air in a calf house, ventilation rate should be 5–6 air changes per hour (q_v = 5–6). If q_d for mixed BCFP in air ranges from 50–100, this means that microbial death is 10–20 times as important as ventilation as a mechanism for clearance of BCFP. For those pathogens for which q_d exceeds 200, the contribution of ventilation to clearance becomes trivial. This point is rather technical but extremely important. It is, in practical terms, impossible to compensate for an increase in stocking density (N/V) by increasing ventilation rate, since stocking rate determines about 90% of the release of BCFP into the air, whereas ventilation may account for only 10% of total clearance of all BCFP, and considerably less than 10% of clearance in the case of pathogens of the respiratory tract. This point is illustrated in Figure 12, which plots concentration of BCFP in a calf house as a function of (i) volume of air per calf (V/N m^3) and (ii) ventilation rate (air changes per hour). This figure is largely self-explanatory. As an illustration, however, examine the dotted line XZY. Increasing the space available to the calf from 5 to 10 m^3 at a ventilation rate of 4 air changes per hour halves the concentration of BCFP in the air from 3060 to 1530/m^3 (XY), for reasons that should be obvious from Equation 4.7. A tenfold increase in ventilation rate from 4 to 40 air changes/h reduces BCFP concentration by only one-third (XZ). A tenfold increase in ventilation rate is, in this example, only 2/3 as effective as a twofold increase in air space per calf in improving air hygiene. We do not know yet just how important air hygiene is in preventing calf pneumonia but we can say for certain that it is impossible to compensate for increasing stocking density by increasing ventilation rate within practical limits. A ventilation rate of 40 per hour cannot be achieved by natural ventilation in still air. It would be prohibitively expensive to create by fan ventilation and in any circumstances it would be almost certain to create unacceptable draughts at calf height.

The environmental factors that determine q_d are, as yet, imperfectly

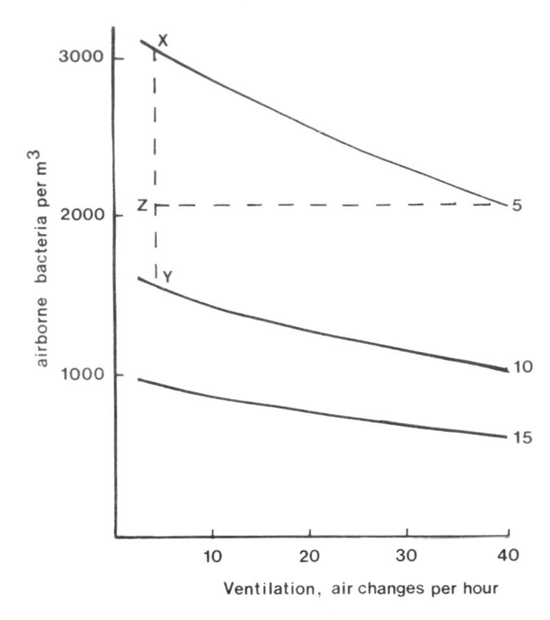

Fig. 12. Effect of space allowance (V/N, m^3 per calf) and ventilation rate (air changes per hour) on the concentration of airborne bacteria in a calf house. The intervals XY and XZ indicate, respectively, the effects of increasing space allowance from 5 to 10 m^3 per calf (XY) and increasing ventilation rate from 4 to 40 air changes per hour (XZ).

understood. It is, however, fairly certain that BCFP die more quickly when relative humidity (RH) is below 80%. Values of RH in excess of 85% are generally considered to be 'unhealthy' for calves. A reasonable explanation for this belief is that high RH prolongs the survival of micro-organisms in the air and on the surfaces of animals and buildings and so increases the magnitude of the challenge to the animals, both by contagion and by airborne infection. One of the prime functions of ventilation is to remove water vapour. The importance of ventilation is therefore potentially greater than that described simply by q_v. If ventilation (and good drainage) reduce RH, they can contribute to air hygiene by increasing q_d. However, in an unheated calf house it is not possible to reduce RH by ventilation below that of the incoming air. During the damp, dark days of winter RH may be above 85% for days on end and in such conditions the concentration of BCFP inside a calf house is likely to be high, however well it is ventilated. In such circumstances the heated, insulated calf house has some merit, since raising the temperature above that outside increases its water holding capacity and so reduces RH. Heating the building may therefore reduce the risk of pneumonia by cleaning the air, rather than by preventing the calves from 'catching cold'. It should be apparent by now that it is of little consequence to a healthy calf whether it is housed at 6 or 16°C.

Control of q_d appears therefore to offer the best hope for improving air hygiene in calf houses. It follows that techniques which remove BCFP by filtration, or kill them, for example by rapid change of RH in an air conditioner, may offer the best hopes for the future for the control of respiratory disease through control of air hygiene.

PENETRATION AND CLEARANCE OF THE RESPIRATORY TRACT

In order to cause disease, invading organisms must first colonise the sensitive tissues of the host animal. In order to cause pneumonia, pathogenic organisms must, by definition, get into the lungs. The anatomy of the respiratory tract of calves and mechanisms of clearance of organisms and other particles deposited in various parts of the plumbing are illustrated in Fig. 13. The first filter for particles in inspired air is provided by the turbinate bones in the nose. This effectively removes all particles greater than 5 μm (0.005 mm) in diameter. In outdoor air or in calf houses at RH less than 70%, about 90% of BCFP are carried on particles greater than 5 μm in diameter. This is because smaller particles dry out so rapidly. When RH exceeds 80%, the number of BCFP carried on particles *less* than 5 μm in diameter may increase to 50% because they survive longer. High RH is therefore doubly harmful. It increases the number of micro-organisms in the air by reducing q_d and it increases the risk that they will be carried on aerosols small enough to penetrate the lower respiratory tract and deposit on the small terminal bronchioles or in the alveoli, or air sacs of the lungs, where gaseous exchange takes place.

The first area of lung reached by incoming particles is the right apical lobe (Fig. 13). The position of this lobe and the angle of contact between its bronchus and the trachea are such that it is likely to catch a high proportion of incoming particles. It is not surprising therefore to discover that this lobe is most commonly affected by pneumonia. The diaphragmatic lobes, which are farthest from incoming particles, are the least likely to become infected.

The lower respiratory tract, from the larynx down, has two mechanisms for clearing infectious organisms or particles in inert material. The walls of the trachea and bronchi contain cells which secrete mucus which traps particles. Other cells possess cilia, small processes which have a wave-like action whose function is to carry mucus and trapped particles (phlegm) up from the bronchioles, along the trachea and to the pharynx, where it is usually swallowed (i.e. removed down the oesophagus). This clearance mechanism is usually given the rather splendid title of 'The Mucociliary Escalator' (Fig. 13).

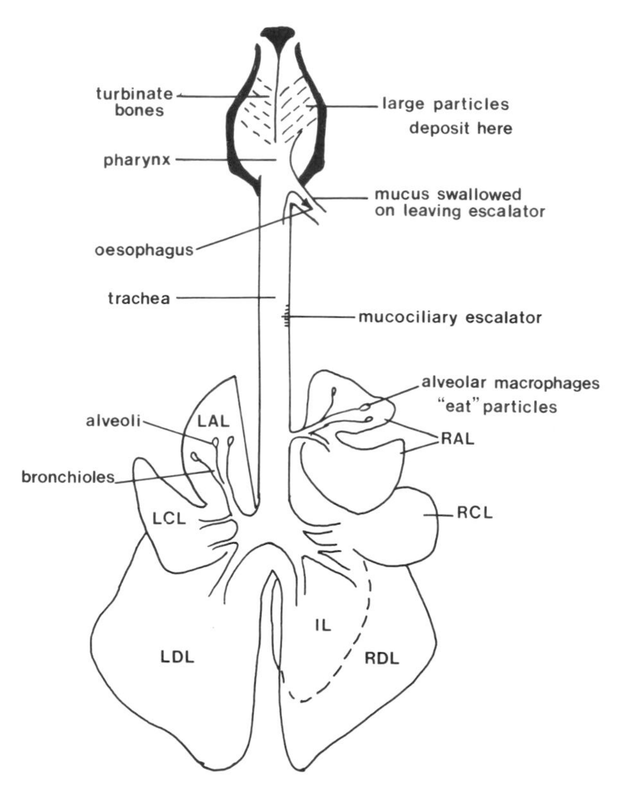

Fig. 13. A schematic illustration of the respiratory tract of the calf, illustrating sites of deposition and clearance of micro-organisms.

The second clearance mechanism involves macrophages or scavenger cells which live in the alveoli and terminal bronchioles and 'eat' incoming particles. That is, in fact, a gross over-simplification. The nature of the battle between micro-organisms and macrophages is extremely complex and quite outside the scope of this book. For a simple introduction to these and other immune mechanisms of defence against infection the reader would do well to consult Playfair

(1979). To illustrate just one of many complications, one property that may cause an organism to be a successful pathogen may be its ability to survive and even reproduce within a macrophage.

It is sufficient for the moment, however, to point out that the lower respiratory tract of healthy calves is being continuously bombarded with a large number of bacteria, dust and other particulate pollutants which need to be removed by the alveolar macrophages and the mucociliary escalator. As indicated earlier, most of the organisms in the air appear to be relatively harmless in themselves, i.e. they do not provoke disease if introduced into the respiratory tract of an otherwise uninfected animal. It is reasonable to suppose, however, that when they are present in very high concentration in the lower respiratory tract they may embarrass or overload these clearance mechanisms to the point where they are not able properly to clear the few genuine pathogens that may be there as well. In these circumstances the balance of infection (Fig. 10) is tipped towards disease.

This theory cannot be proved from existing evidence but it is consistent with the clinical picture of respiratory disease in calves. The incidence of pneumonia and other coughing diseases severe enough to merit treatment is much higher in bought-in calves than in those which are home-reared. The disease, however, may not appear for several weeks after arrival of the calves at the rearing unit, although potentially infectious organisms like *P. haemolytica* may be isolated from the nose and pharynx at any time after arrival. The greatest incidence of calf pneumonia occurs between 6 and 10 weeks of age. Various factors unrelated to housing, such as possible stresses at weaning, and loss of passive immunity from colostrum, may contribute to this clinical picture, but there is also evidence that air hygiene tends to be worst in calf-rearing houses, or 'follow-on' houses at about this time. Bad air hygiene, by overloading the clearance mechanisms in the lungs, may create an environment wherein pathogenic organisms acquired weeks earlier can now multiply and cause damage to the bronchioles and lungs. Once the damage has started, the airways get obstructed, mucociliary clearance is impaired and things go from bad to worse until or unless the immune mechanisms of the body can mount their own defence. It seems sensible to assume also that gaseous air pollutants like ammonia or sulphur dioxide make it easier for infectious pathogens to get a hold by damaging cell surfaces or slowing down the escalator. In fact, there is no good evidence that at levels commonly encountered in animal houses these gases are doing any serious harm, but this lack of evidence should not be taken as a justification for foul-smelling calf houses. A trained nose is still one of the best indicators of good air hygiene in a calf house. If it doesn't smell good, it isn't.

SYSTEMIC RESISTANCE TO INFECTION

Any severe environmental stress such as heat, cold, exercise or fear can reduce the ability of an animal to mount a defence against infection. Stresses of this nature are usually considered to fall into three stages.

(1) The first phase of the response to any sudden change in the environment is known as the *alarm reaction*. During the alarm stage of the response to a sudden stress such as a blizzard, or some rough handling during loading onto a lorry, the immune mechanisms involved in resistance to disease are usually suppressed. This immune suppression does not outlast the duration of the stress, but undoubtedly there is an increased risk that an infection may 'take a hold' during this time.

(2) If the stress is prolonged (e.g. cold stress) the animal enters the *adaptation* phase. Here it has come to terms with the environmental stress, albeit at some physiological cost. The nature of the immune response during adaptation to different stresses has been reviewed by Kelley (1982). The subject is bewilderingly complex, but as a general rule, it is just about fair to say that during the adaptation period the ability of the animal to mobilise an immune defence against infection is almost as good as in the absence of stress. In some cases it may be better. Adaptation to mild cold stress may even enhance the ability of an animal to resist infection by *Pasteurella haemolytica*.

(3) Stress cannot be borne for ever. Either the animal adapts to the point that the stress disappears or it ultimately reaches the stage of *exhaustion*. Cattle have an enormous ability to adapt to stresses of both cold and heat. Equally, they can adapt to several stresses involved, for example, in mixing with new groups. It is not too far from a truism to say that the optimal environment (in any sense) for a cattle beast is the one that it has been in for the last three weeks. Nevertheless, some stresses on calves can be so severe and so prolonged that the animal can reach the point of exhaustion. At this point, all its defence mechanisms are impaired. Movement of

the dairy calf in the first few days of its life is potentially the most damaging series of events that it will ever experience because it combines the greatest possibility for infection with the greatest possibility of severe damage to its mechanisms for resisting infection.

The transport of calves need not be a stressful experience. There is good evidence that calves are not frightened by transport and mixing as such. This is hardly surprising since a newborn infant simply has not learnt the meaning of fear. Young calves are more *bewildered* than anything else by the things done to them in the first days of life. They do, however, appreciate the need for both food and warmth and if this is denied them while they are moved around on lorries and through markets they very quickly become distressed and fairly quickly become exhausted. Management of the animal from market will be covered in Chapter 9.

Summary of environmental needs

AIR TEMPERATURE

Air temperature is unimportant for normal, healthy calves provided they are kept dry and out of draughts. The sick calf or the calf that has been starved and stressed during transport may, during cold weather, benefit from some supplementary heat, perhaps provided by an infra-red lamp over its pen.

AIR MOVEMENT

Air movement at calf height should not exceed 0.25 m/sec (50 ft/min) during winter conditions to avoid the risk of chilling draughts. In hot weather, increased air movement around the animals will, of course, help to keep them cool.

SPACE REQUIREMENT

Recommendations for floor space were given on p. 82. The amount of air space necessary to ensure adequate air hygiene is unknown. Mitchell recommends not less than 6 m^3 (212 ft^3) per calf. This is probably adequate for calves up to weaning at 6 weeks of age (or 60 kg). For rearing calves up to 12 weeks of age I recommend not less than 10 m^3 per calf, and for veal calves raised to 200 kg weight at 16 weeks of age not less than 15 m^3 per calf. These figures are higher than those recommended by many advisers but that is because ventilation is not as effective as we thought it was.

RELATIVE HUMIDITY

Ideally, relative humidity should be kept below 75% to ensure adequate air hygiene. Heating the building to reduce RH is not an economic proposition so there are occasions, especially in winter, when this is not possible. Nevertheless, every attempt should be made to keep RH down by ensuring adequate ventilation, good drainage (1:20 floor slope to efficient drains) and by sparing use of the hose during the time that the calves are in the building.

VENTILATION

A calf house should be well enough ventilated to remove excess heat in summer and in winter to remove carbon dioxide, water vapour and smells. Its contribution to ensuring air hygiene is rather small. Recommended summer and winter limits are given below.

	Winter	*Summer*
Ventilation rate m^3/h per calf	40	100
Air changes/h when $V/N = 10$	4	11

These rates can be achieved either by fans or by natural ventilation. This will be described in the next chapter.

Further reading

Brambell, F.W.R. (1965) 'Report of the technical committee to enquire into the welfare of animals kept under intensive livestock husbandry systems.' Cmd. 2836. H.M. Stationery Office, London.

Bruce, J.M. (1979) 'Heat loss from animals to floors.' Farm Buildings Progress **55**: Scottish Farm Buildings Investigation Unit pp. 1–4.

Clark, J.A. (1981) *Environmental Aspects of Housing for Animal Production.* Butterworths, London.

Hafez, E.S.E. (1968) *Adaptation of Domestic Animals.* Lea & Febiger, Philadelphia.

Kelley, K.W. (1982) 'Immunobiology of domestic animals as affected by hot and cold weather.' Proceedings of 2nd International Livestock Environment Symposium, Ames, Iowa, U.S.A.

Mitchell, C.D. (1978) *Calf Housing Handbook.* Scottish Farm Buildings Investigation Unit.

Monteith, J. & Mount, L.E. (1974) *Heat Loss from Animals and Man.* Butterworths, London.

Playfair, J.H.L. (1979) *Immunology at a Glance.* Blackwell, Oxford.

Sainsbury, D.B. & Sainsbury, P. (1979) *Livestock Health and Housing.* Bailliere Tindall, London.

Webster, A.J.F. (1971) 'Prediction of heat losses from cattle exposed to cold outdoor environments.' Journal of Applied Physiology, **30**: pp. 684–690.

5 Housing design, ventilation and disinfection

A good calf house is one that meets both the needs of the animals and of the stockman (him or her) at a reasonable cost. The husbandry, health and welfare needs of the animal were described in the previous chapter. What the stockman needs is a building that is comfortable and convenient to work in. This obviously applies to routine daily chores like feeding and cleaning out, but the stockman should also be able to observe clearly all calves at all times and to catch, single-handed, restrain and administer medicines to any individual animal requiring attention. Finally, the building and fittings should be designed in such a way that they can easily be cleaned and disinfected between batches of calves.

I am not an architect or buildings designer and this chapter will not tell you very much about how to build a calf house. What it will do is consider a variety of simple and complex forms of calf housing in the context of the needs of the calf and the stockman. Each individual who wants to erect a new calf house or alter existing accommodation can use this information to draw up specifications to suit his unique set of requirements. Those who require further information on calf house design, ventilation and disinfection should refer to Mitchell's *Calf Housing Handbook* (1978), Sainsbury and Sainsbury *Livestock Health and Housing* (1979) and regular publications from the Land Service of the Ministry of Agriculture, Fisheries and Food.

Possible forms of accommodation

THE CALF HUTCH

The environmental needs of the calf are simple: physical comfort, good hygiene and reasonable shelter. The healthy, home-reared calf, eating and drinking normally, will not be stressed by low temperatures as such. Figure 14(a) illustrates a simple type of calf

Fig. 14. Calf hutches. (*a*) Isolated hutches in north/central USA, (*b*) Rows of hutches in Holland.

hutch used in Wisconsin, U.S.A., where winter temperatures regularly fall below −20°C. Given a good bed of straw inside the hutch the calf can shelter from the worst excesses of the weather. The small run at the front allows it to see and hear what is going on but the pens are sited such that it is absolutely impossible for the calves to infect one another by direct contagion and almost impossible for them to transmit infection by the aerial route. These hutches are particularly suited to low rainfall areas like the American Middle West, or the east coast of Scotland. On poorly drained land in the wet dairy areas of western Britain, like Cheshire and Somerset, they could become islands in a swamp.

Being light, the huts can and should be moved between batches of calves to prevent build-up of infection at any one site. It is also necessary to clean and disinfect the individual wooden boxes. Methods of disinfection will be considered later in the chapter.

In the right environment and on the right ground these hutches provide excellent, healthy, cheap accommodation for home-reared calves in the first weeks of life. They would be a little Spartan for bought-in calves in the first two or three days after arrival but not intolerably so. They have one enormous disadvantage. It is difficult to imagine a housing system less well suited to the needs of the stockman, who has to carry all the food (including warm milk) considerable distances between calves, often in most unpleasant weather. Moreover, stockmen whose backs are less supple than they were do not exactly cherish the prospect of catching and restraining for treatment individual calves which can retreat into a hutch four feet high. It is perhaps significant that most of these calf hutches can be found in colleges of agriculture and veterinary science where there are people who preach the concepts of good hygiene and *other* people who do the daily chores.

An alternative form of low-cost calf housing is the row of hutches shown in Fig. 14(b). This is better suited to the wet, windy calf-rearing areas of western Britain because the floors of the individual pens are raised off the ground and the legs are standing on concrete. Needless to say, this is only a low-cost alternative if a concrete standing is already available. On the whole I do not find it particularly attractive. The arrangement for bucket-feeding is awkward and there is little space to let in fresh air and sunlight. That is not a sentimental criticism. Sunlight (even when diffused through cloud) is one of the best possible destroyers of the micro-organisms that accumulate in calf pens. In certain weather conditions, typically when a period of warm, moist weather follows a cold spell, the walls and roofs of these pens would run with condensation. Finally, as Fig. 14(b) illustrates, rows of hutches cannot guarantee freedom from spread of infection by contagion or short-range airborne transmission.

EXPLOITING THE DUTCH BARN

Perhaps the most attractive of all forms of simple calf shelter is that which can be created under the high roof of a Dutch barn or similar multi-purpose covered area. Figures 15 and 16 illustrate two variations on this theme. In Fig. 15, straw bales have been used to provide individual calves with shelter from cross-winds and down-draughts. As the calves grow, the straw bales which have been used to provide a partial roof over the pens can be broken out and added to the bedding. Figure 16 illustrates a high-roofed barn with permanent walls but ample ventilation above calf height.

Here the calves are again provided with deep straw bedding and excellent shelter from all draughts.

The advantages of setting up the individual sheltered pens under a high roof are enormous. For a start it keeps all, or nearly all, the rain out. It is practically impossible to keep snow out of any large, airy farm building but, as a general rule, snow tends to get sucked in longer distances through narrow gaps than wide ones and provided that calves can get under the half-roofed area of their individual pens, they are unlikely to suffer much. Keeping the rain out not only keeps the calves and the bedding dry but also provides a better working environment for the stockmen and also enables him to store food and bedding in easy proximity to the animals. The final advantage of the high roof is that it provides a very effective barrier against condensation, acting in much the same way as does an awning erected over a tent. Either arrangement (Fig. 15 or Fig. 16) provides for all the reasonable thermal needs of the calf and also ensures very respectable standards of air hygiene, since the allocation of air space to each calf is enormous in comparison to that existing in most custom-built calf houses.

Each of the illustrations shows open-sided pens. These are perfectly satisfactory for home-reared calves which may be presumed to be healthy on arrival in the unit or at least carrying only those germs common to the farm. A farmer rearing bought-in calves might

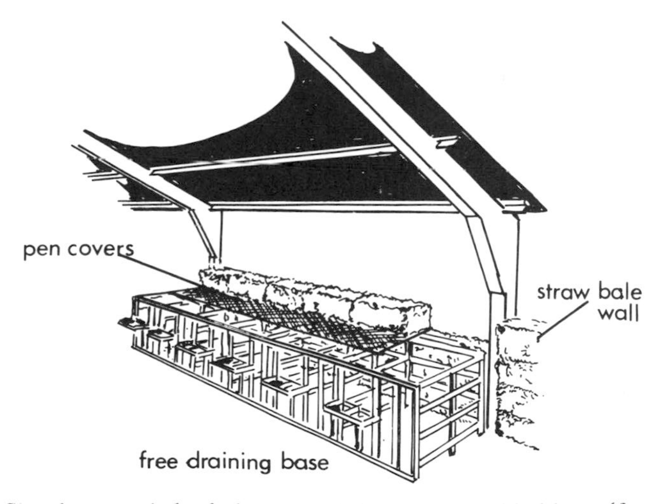

Fig. 15. Simple straw-bale shelters erected in an open-sided barn (from Mitchell, 1978).

Fig. 16. Calf kennels in a Dutch barn.

consider it worth spending considerably more money and constructing solid-sided pens (see Fig. 5(a)) to restrict contagion. Because all but the most virulent enteric infections can usually be kept in check by careful control of feeding in the first few days after arrival (Chapter 9), I do not personally consider solid-sided pens to be essential.

It is also possible to rear calves in groups under the roof of a Dutch barn. Here again, there should be an area covered by a simple straw-covered roof that all the calves in the group can get under when they wish. If new-born calves are brought into the unit when the weather is very cold, I like to hang infra-red lamps under these roofs, at a safe distance from both the calves and the straw. One lamp will suffice for about six calves, two or three for a pen of about sixteen animals. It may be necessary to switch these on only at night and, provided the calves remain healthy, they can be removed after a week.

Feeding cold acid milk through teats to calves reared in a Dutch barn can be a problem in cold weather. I do not believe that cold milk stresses the calves when it is provided free-choice. In cold weather calves tend to consume the same amount per day but in smaller amounts at any one meal and the effect on their total heat loss is quite small. The main practical problem is that the milk freezes up. This is one case where an automatic dispenser of milk at blood heat really comes into its own, since it is a far cheaper option than erecting a specialist calf house and probably a healthier one, too. It is important here to ensure that the teats are kept as close to the machine as possible to keep the milk warm. A machine capable of dispensing to 30–40 calves would ideally be situated at a well-drained site between two pens to provide two teats for each lot of 15–20 calves.

MONOPITCH CALF HOUSE

Mitchell (1978) has designed a number of extremely successful specialist calf-rearing houses of monopitch construction. One of these is illustrated in Fig. 17. In essence these buildings are no more

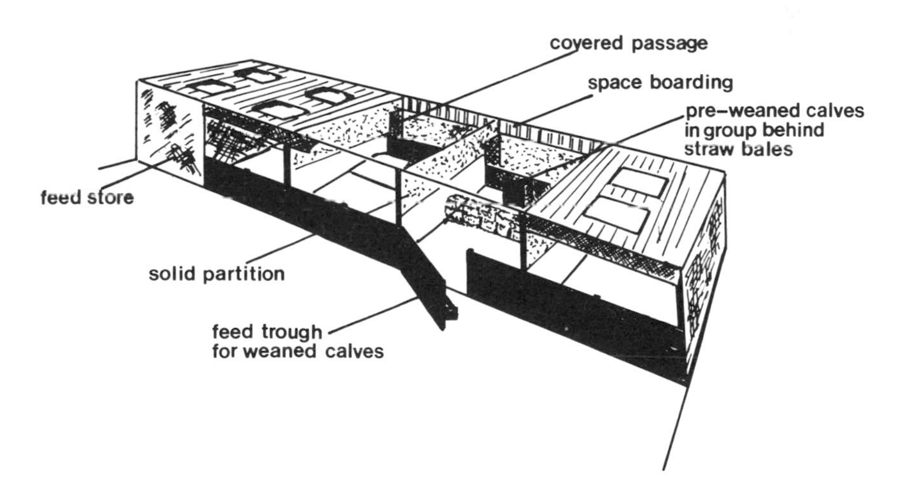

Fig. 17. A calf house of monopitch design (adapted from Mitchell, 1978).

than well-built follow-on houses designed to provide calves with shelter but plenty of fresh air. In fact, there is quite a strong argument in support of the case that the best form of calf house is a good follow-on house. The Mitchell monopitch is a reasonably low cost affair and has several features which make it particularly attractive in the context of calf health and husbandry.

(1) The partitions between bays are solid to roof height. This prevents draughts developing down the length of the building and restricts airborne spread of infection from one bay to another. The cement-plastering at calf level ensures ease of cleaning between batches.
(2) His insistence on a 1 in 20 slope on all floors ensures good drainage and improves hygiene both on the ground and in the air since good drainage helps to keep relative humidity down. It also cuts down on the amount of straw required.
(3) Space boarding at the back of the building ensures good ventilation and mixing of air in all areas. The down-hung sheeting at the front greatly restricts ingress of rain and snow.

Mitchell's (1978) original drawings illustrate the use of 'tethered feed fences' for calves in the first weeks of life. These are alternatives to solid-sided pens for individual calves. They are, of course, cheaper because the calves are tethered and there is no need to extend the sides of the pens any further back or forwards than the calf can stretch its neck. They are hygienic and calves appear to settle well into them but I must say that I prefer an individual pen that allows the calf complete freedom of movement, particularly in an unheated, uninsulated calf house where a slight adjustment of position, such as curling up under the half-roofed area of the pens shown in Fig. 16, may make a considerable difference to comfort.

Another bay is shown (Fig. 17) to contain an automatic dispenser for feeding calves in groups. When the calves are small, they can be restricted to the rear of the bay by straw bales. These provide shelter and a small roofed section would again be a good idea in midwinter. Any rain or snow will enter the front of the bay, away from the animals. As they grow, this shelter can be broken out and the calves given access to the entire area. Each bay of the standard Mitchell monopitch building has a volume of 160 m^3. Three rows of tethered feed fences can accommodate 24 calves before weaning, giving an air space of $6\frac{2}{3}$ m^3 per calf. This is quite adequate for the preweaned calf. If calves are to be kept in this building for the whole of the first winter I would like ideally to see 16 and at most no more than 20 in each bay, giving 10 or 8 m^3 air space per calf respectively.

The open end of each bay should face south (in the UK). This allows maximum opportunity for winter sun to enter the building, warm the calves and clean the air. In summer when the sun is higher in the sky, the roof provides shelter from excess heat. The size of the openings to the building relative to its dimensions are such that there is good mixing of air without draughts at calf height and ventilation rate in still air conditions out of doors is at least six air changes per hour. Provided that the building is not overstocked, the air is respectably fresh and the incidence of respiratory disease in these units is usually very low. Empty bays can also be put to a variety of other farm uses. On most monopitch units there is a corridor at the north, enclosed, back end of the building to give the stockman access under cover to calves and feeders in each bay. Curiously, this is the coldest and draughtiest part of the building, but it is at least dry. All in all, the Mitchell monopitch must be considered a very successful compromise between the needs of the calf for space and fresh air and the need to provide reasonable working conditions for the stockman.

VENTILATION OF THE ENCLOSED CALF HOUSE

Fully enclosed calf houses range from custom-built, insulated, fan-ventilated new units to conversions of old stone farm buildings with slate roofs or (less successful) conversions of not-so-old wooden chicken houses.

Since there is now general recognition that fresh air is more important than warmth to the young calf, the first thought in converting an old building or designing a new one for calves is to ensure adequate ventilation. The limits of what can be achieved by ventilation were described in the previous chapter. Ventilation removes heat, moisture and chemical substances. It also removes some of the micro-organisms in the air, but the number removed by ventilation is far less than the number that die in the building. In summer, dissipation of heat assumes major importance and maximum ventilation rates must take this into account (110 m^3/h per calf or 11 air changes per hour when air space per calf is 10 m^3). In winter the main function of ventilation is to remove moisture from the building and keep relative humidity down, ideally below 80%, so as to reduce the survival time in the air of small aerosols containing micro-organisms.

Sainsbury and Sainsbury (1979) provide a detailed description of the role of ventilation in removing heat and moisture from livestock buildings and it is not necessary to reproduce here all the mathematics involved in these calculations. Two equations only are necessary to convey a general understanding of what ventilation does.

Heat is lost from an animal house by two routes:

(i) 'Building heat loss' (BHL) describes heat loss through the structures of a building at a rate determined by the dimensions of the building and their thermal insulation. BHL is expressed in watts/°C temperature difference between inside and out.
(ii) 'Ventilation heat loss' (VHL) describes heat loss from the building in outgoing air. This depends on ventilation rate and the specific heat of air. A ventilation rate of 1 m^3/h corresponds approximately to a VHL of 0.336 watts/°C.

The temperature rise (°C) incurred as a result of the heat generated by calves within a building is given by

$$\text{Temperature rise (°C)} = \frac{\text{Heat production by calves (watts)}}{\text{BHL + VHL (watts/°C)}} \quad (5.1)$$

The higher the temperature of the air in a building, the more moisture it can carry. For example, saturated air at 5°C contains 5.7 kg H_2O/m^3. Expressed in another way, the absolute humidity of air is 5.7 kg H_2O/m^3 at 5°C when relative humidity (RH) is 100%. The absolute humidity of saturated air at 15°C is 12.4 kg H_2O/m^3. If one simply added heat to the air to raise the temperature from 5 to 15°C it would reduce relative humidity (RH) from 100% to 46% (5.7/12.4). There are those who believe that calves are less likely to 'catch a chill' (i.e. suffer from respiratory disease) in a heated calf house. I used not to subscribe to this belief since demonstrating just how trivial is the direct stress of cold inside the average unheated house. However, supplementary heat in a calf house may reduce the risk of respiratory disease indirectly by lowering RH and so improving air hygiene, but it does seem to be a terribly complicated and expensive alternative to giving the calves reasonable access to fresh air.

In a cool environment (say at 10°C), calves lose about 20–25% of their heat by evaporation. The evaporation of moisture from the skin surface and respiratory tract of calves and from building surfaces increases the absolute humidity of the air. To keep RH down to an acceptable level, ideally below 80%, it is necessary (a) to ensure a temperature rise in the air as it enters the building so as to increase its water-carrying capacity, and (b) to ventilate the building so as to carry moisture out. It is obvious that there is some conflict between these two objectives since the greater the ventilation rate the lower the rise in air temperature.

The effect of ventilation on RH (%) inside an animal house can be calculated from the following equation:

$$\text{RH}(\%) = \frac{100\,(W_a/Q + q_o)}{q_i\ \text{sat}\ (T_b)} \qquad (5.2)$$

where q_i is absolute humidity inside the building (g/m³)
q_i sat (T_b) is absolute humidity of saturated air at building temperature (g/m³)
q_o is absolute humidity of outside air (g/m³)
W_a is moisture production from animals and building surfaces (g/h)
Q is ventilation rate (m³/h)

Consider an insulated building (say a converted poultry house) containing 25 4-week-old calves in an air space of 150 m³. This provides 6 m³ per calf.

Heat loss from calves	= 4.2 kw
Moisture loss from calves	= 1.5 kg/h
from building	= 0.5 kg/h
Outside air temperature	= 5°C, RH = 80%
Ventilation rate	= 4 air changes/h = 600 m³/h
Building heat loss	= 0.30 kw/°C
Ventilation heat loss = 0.336 x 600 x 10^{-3}	= 0.20 kw/°C
Temperature rise in building (Eqn. 5.1)	= 8.4°C
Inside air temperature	= 13.4°C
Internal RH (Eqn. 5.2)	= 70%

In this example, a ventilation rate of four air changes per hour in an insulated building has achieved a temperature rise in the building from 5 to 13.4°C and a fall in RH from 80 to 70%. Figure 18 illustrates the effects of varying ventilation rate on RH in a building with and without insulation. At a ventilation rate of one air change per hour (150 m^3/h) the building will be saturated with water vapour and condensation will form on the roof and walls. The greatest reduction in RH is achieved at 8 air changes per hour in the insulated building and at about 14 air changes per hour in the uninsulated building. At ventilation rates in excess of this, RH increases slightly as the temperature rise in the building gets smaller and smaller. It has been shown already that the most important contribution of ventilation to air hygiene is its role in reducing RH. Figure 18 suggests that 8 air changes an hour in an insulated building well stocked with calves is sufficient to keep RH below 80% even in cool, damp conditions out of doors (air temperature 10°C, RH 90%). In this example RH never falls below 80% in the uninsulated building whatever the ventilation rate. Some insulation may be indicated for calf houses, not so much to keep the calves warm as to keep RH down and so help to preserve reasonable standards of air hygiene.

The example shown in Fig. 18 assumes a moisture load on the air from urine, faeces and water released into the building of only 0.5 kg/hour. In some circumstances it can be far greater than this, particularly if the calf house is frequently hosed down. This presents

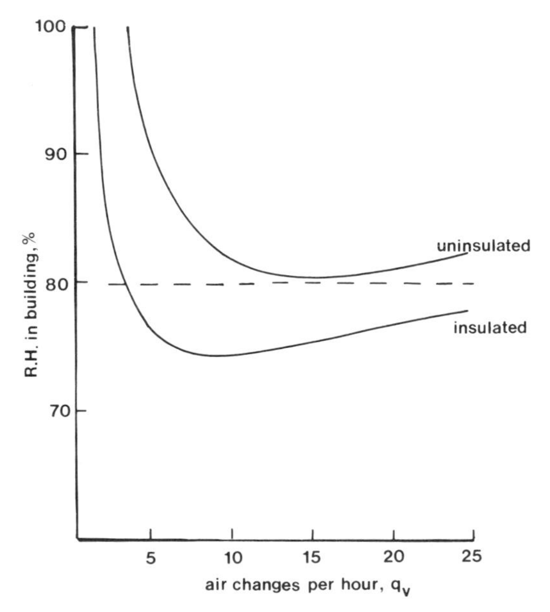

Fig. 18. Effect of increasing ventilation rate on relative humidity (RH) in a fully-stocked calf house when air temperature = 10°C and RH outside is 90%. The two curves illustrate the effects in an insulated and an uninsulated building.

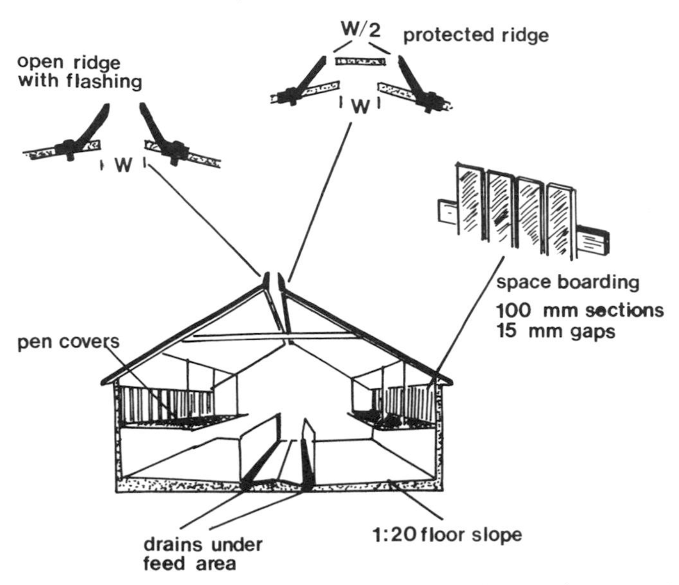

Fig. 19. Specifications for a naturally ventilated calf house (adapted from Mitchell, 1978). Minimum allowance for adequate natural ventilation are (*a*) eaves inlet = 0.05 m^2 per calf, (*b*) ridge outlet = 0.04 m^2 per calf, (*c*) height between inlet and outlet 1.5 m.

the stockman with a dilemma; he wishes to use the hose to keep the surfaces clean but needs to restrict the water load on the building to keep the *air* clean. The best solution to this problem is to ensure first-class drainage which can do almost as much as good ventilation to keep down RH.

NATURAL VENTILATION

Natural ventilation means the movement of air through a building brought about by natural forces acting in the air itself. On a windy day these forces are fairly obvious. However, it is possible by appropriate design to achieve eight air changes per hour in a calf house even when outside air is absolutely still. The heat generated by the calves themselves causes the air within the building to rise. An opening placed at a high point in the building (e.g. the ridge of the roof, Fig. 19) will act as an outlet for stale air and other openings placed lower down (e.g. at the eaves) will act as inlets. This 'stack effect' only works if there is a sufficient difference in height between the openings (not less than 1.5 m, preferably 2.5 m).

The example shown in Fig. 19 has space boarding at the eaves. The minimum open area at the eaves should be 0.05 m^2 per calf.

The space between the 100 mm boards should be 15 mm. This allows sufficient air into the building on still days but prevents severe draughts when a gale is blowing outside. Heavy duty plastic netting or even chicken wire can be used at a pinch as an alternative to space boarding.

The ridge should provide 0.04 m^2 open area per calf. The example shown in Fig 19 is an open ridge with apron flashing pieces as upstands to each side. When the building is full of calves the warmed air leaving by way of the open ridge keeps out nearly all the rain in still air conditions. (If the building is empty the rain does get in!) In conditions of driving wind and rain the flashing pieces keep water from blowing up the roof and falling into the building through the open ridge. The small amount of rain that may enter falls into the centre passage rather than onto the calf pens.

Most cattle houses that still possess a capped, or covered ridge, do not provide anything like 0.04 m^2 per calf. If a protected ridge is to be provided it is essential to ensure that this minimum area is achieved. An effective weather-proof example suitable for use where the ridge may extend over part of the calf pens is also shown in Fig. 17.

Suspended below the eaves are solid pen covers to prevent down-draughts of cold air falling directly onto the calves. It would be equally satisfactory to roof over the backs of the individual pens with straw and wire as shown in Fig. 16. At a stocking rate of 6 m^3 per calf this would be a very good house, providing a satisfactory balance between comfort and fresh air. It cannot be emphasised too strongly that *no attempt should be made to block up any of the air inlets or outlets, even in the most severe weather.*

According to Mitchell, there is no need to insulate the roofs of naturally ventilated calf houses. He is probably correct that it is not particularly cost-effective since the calves do not require additional heat and at a ventilation rate of not less than eight air changes per hour the temperature inside the building is little greater than that outside, so building heat loss is small. Nevertheless, there are some advantages of an insulated roof, or at least a double-skinned roof. As indicated earlier, the insulation or outer skin cuts down radiant heat loss from the inner skin and thus prevents its temperature falling below air temperature in the building. This reduces condensation and also ensures that warmed air rising from the calves does not cool down too much so that it passes through the ridge rather than falling back on the calves. An insulated roof in a naturally ventilated building will therefore both improve ventilation and reduce RH, both of which are highly desirable (see Fig. 18).

There are two limitations to a naturally ventilated, unheated animal building. In still air conditions in high summer, natural ventilation cannot achieve the ventilation rate necessary to keep air temperature down to acceptable limits. This can be a problem for pig and poultry houses. Most calves would expect to be out-of-doors in high summer but veal calves in naturally ventilated houses can experience distress in very hot conditions.

The second limitation is more serious. In the cool, damp, sunless days of winter when outdoor air temperature is, say, 5°C and relative humidity is near to 100%, it is impossible to keep RH inside the building below 85% by ventilation alone. In these circumstances the calf's risk of contracting pneumonia is increased and there is little that the stockman or veterinary surgeon can do about it without spending a great deal of money to install air conditioning.

FAN-VENTILATED BUILDINGS

Since it is possible to achieve a satisfactory ventilation rate by natural means there is no reason to consider fan ventilation when designing a building from scratch since fans are expensive to install, they require maintenance and there is always the risk of power failure. Fan ventilation may, however, be indicated where calves have been, or are to be put in a building that was not designed for them and there is no easy way of providing the building with effective natural ventilation, perhaps because it has a flat roof so a stack effect becomes impossible to achieve, or because it has stone walls two feet thick. In these circumstances it pays to seek professional advice. Sometimes things are not as bad as they seem. An uninsulated slate roof on an old farm building (a 'breathing roof') may allow a lot of air to escape between the slates, thereby reducing or even removing the need for extractor fans.

The specifications for a fan-ventilated calf house are the same as before: not less than six air changes per hour in winter conditions, achieved in such a way as to avoid direct draughts on the calves. Maximum ventilation rates for warm conditions should be 11 air changes per hour. The fans should be left to run continuously. If fan speed is regulated by a thermostat it must be arranged so that fan speed never falls below that required to sustain six air changes per hour. *The calf house fan that switches itself off when the weather gets cold is a potential death trap.*

CONTROLLED ENVIRONMENT

From time to time stock farmers in the developed world have taken advantage of the occasional good year and invested in controlled environment calf houses. A large number of these were built in the 1950s and 1960s at about the same time as controlled environment buildings were being successfully developed for pigs and poultry. In this case, the expression 'controlled environment' really meant controlled air temperature and the objective was to improve efficiency by improving food conversion ratio. The 1960 generation of controlled environment calf houses were designed with the best of intentions to achieve the same initial objective, namely an air temperature of not less than 12°C (54°F), although there was sufficient evidence at the time to indicate that any benefits in terms of feed conversion ratio were likely to be trivial. The buildings were insulated and both fans and heaters were operated thermostatically. In order to keep running costs down, stocking rate was very high (4 m^3 per calf or less) and when outside temperature fell the fans would either turn off or reduce ventilation rate to less than two air changes per hour. In view of what has been said so far it is not surprising that these 'controlled environment' calf houses soon acquired a reputation for doing more harm than good. It would, however, be unfair to condemn the concept of environmental control out of hand because the designers of the 1960s were given the wrong specifications.

There are circumstances where it may pay to invest a considerable amount of capital to build a new custom-designed controlled environment house for calves; for example, on a specialist unit rearing bought-in calves for 30 days before export from the U.K. to the continent of Europe. In these circumstances the architect is given the following specifications:

(1) Air space per calf: 5 m^3 minimum
(2) Air temperature, acceptable range: 10–20°C
(3) Relative humidity, acceptable range: 60–75%
(4) Air hygiene: not greater than 1000 BCFP/m^3 air

These would present him with a few problems, since they could not be guaranteed at a stocking rate of 5 m^3 per calf by heating and ventilation alone. Expensive air-conditioning would be required to achieve specifications 2 and 3. The final objective would require the installation of some form of filter or scrubber to remove BCFP from the air within the building. If this could be achieved then

5 m^3 per calf would be ample since the only point of recommending a greater space allowance was to improve air hygiene.

THE FOLLOW-ON HOUSE

The home-reared calf weaned properly at 5–6 weeks of age and weighing about 60 kg (for a Friesian) is potentially a strong, cold-hardy individual able to thrive in comfort with a minimum of shelter. Nevertheless it is more likely to suffer from pneumonia at this time than at any other. The reasons for this are not surprising. By 6–8 weeks of age, passive immunity is very low. If calves reared in individual pens are then mixed for the first time with each other and with older cattle, they are likely to experience infections at a time when their resistance is at its lowest ebb. The best form of follow-on house is therefore one that is least likely to expose calves to new infections. If possible, recently weaned calves should not be turned into large barns with older cattle, even though these older animals are showing no signs of sickness. Ideally, calves from the dairy herd should be kept in the same relatively small groups until they go out to grass for the first time. I made earlier the somewhat provocative remark that the best sort of calf house is a good follow-on house. This works both ways. If calves can be reared to weaning in a building that can also be used as follow-on accommodation, they avoid the stresses of mixing after weaning and are likely to show healthy, uninterrupted growth throughout their first winter. The monopitch unit illustrated in Fig. 17 is ideal for this approach. If I wanted to grow dairy heifers fast with the intention of calving them down at two years of age, I would rear them in groups on *ad lib.* milk, wean them at 8 weeks of age (see Fig. 6) and keep them in the same bay of a monopitch building until turn-out in the spring. It would not be the cheapest approach in the short term but it would be likely to pay off in the long run.

CLEANING AND DISINFECTION

There is absolutely no doubt at all that if you want to rear healthy calves it pays to start them off in a building that is as clean as you can possibly make it. At first sight this seems obvious, at second sight perhaps a little surprising, since one would expect the risk of calves acquiring infection by direct contagion from another animal or by the airborne route to be greater than the risk of acquiring infection off the surfaces of buildings. Nevertheless, there is clear evidence that buildings get 'calf sick' when repeated batches of animals are run through the building, i.e. when all other factors are taken into account, there remains a tendency for disease incidence to increase with successive batches, and liveweight gains to diminish. At third sight, therefore, one may suppose that if even small numbers of disease-causing organisms are present in a building when new calves arrive, their population (and thus the infective dose) will start to build up at a greater rate than the calves can develop their immunity, and disease will supervene. If the building is clean on arrival, infection and immunity can be kept in balance (see Fig. 10).

There are three essential steps in preparing a calf house for its new arrivals, cleaning, disinfection and rest. *Cleaning* means the removal of micro-organisms and the dirt (dust, manure, etc.) that provides them with food and shelter. *Disinfection* means the killing of micro-organisms, bacteria, viruses and fungi by physical or chemical means. Strictly speaking, a *disinfectant* is a substance used to kill organisms on inanimate things like the surfaces of buildings; an *antiseptic* is a substance used to kill organisms on living tissues like the skin. It follows, therefore, that disinfectants tend to be toxic to humans or calves by virtue of their chemical composition or simply the concentration in which they are applied and should be used with care and attention to the manufacturers' instructions. *Fumigation* is a specialised form of chemical disinfection using a vapour rather than a liquid. The third member of the trio, *rest,* means rest *after* cleaning and disinfection. A building is not being properly rested if it is still full of manure or faeces-encrusted calf pens. The purpose of rest after cleaning and disinfection is to allow naturally occurring disinfection processes like sunlight and desiccation to finish off most of the few remaining micro-organisms.

Cleaning

Ideally, all internal fixtures like calf pens should be dismantled and, if possible, removed from the house. A large tank is useful, first for soaking the sections of the pens in water to soften and remove dung and other debris. The sections should then be removed from the tank and scrubbed down, power-hosed or steam-cleaned. The tank can be refilled with a solution of disinfectant and the clean sections re-immersed. Finally, they should be left to dry, preferably in direct sunlight.

The Ministry of Agriculture (Leaflet 645, 1981) wisely recommend that before starting to clean the house it is most important to ensure that the electricity is off and remains off while the building is wet.

They further advise that it is a wise precaution to remove all fuses.

All bedding and other solid material should be removed from the building and then all surfaces cleaned as well as possible. All surfaces includes things like joists, ledges and window panes which accumulate considerable amounts of dust. If it is necessary to use a bucket of water and a scrubbing brush (and I would rather hope it wasn't) then add washing soda (sodium carbonate) to the water at 1 kg in 25 ℓ (1 lb in 2½ gallons) or a non-foaming detergent as directed on the bottle. Gloves and goggles should be worn.

A power-hose is far more effective at dislodging muck from surfaces when the surfaces are of rendered concrete or some other non-porous material, and drainage is good. It is not a good thing to turn a power-hose onto an insulated roof or electrical trunking. An industrial vacuum cleaner is better at getting dry dust from high level fixtures and fittings.

Steam-cleaners are expensive but very effective. The sort that produces true steam is doubly effective both as a cleaner and as a disinfectant since it kills nearly all micro-organisms. Hot water, high pressure washers generate a greater force of water at about 94°C. This probably removes more dirt and bacteria than a true steam-cleaner but is less likely to kill them *in situ*. Either sort of machine works very well provided the surfaces of the building can stand the assault. If you are only putting through two batches of calves a year it may be better to hire one rather than buy one.

Cleaning is undoubtedly the most important single stage of the preparation of a calf house not only because it removes the vast majority of the disease-causing organisms, but because it removes the shelter of organic material from those remaining and so exposes them to the action of natural and artificial disinfectants.

Disinfectants

For purely practical purposes the liquid chemical disinfectants can be considered in two groups, the phenols and all the rest. The phenols or carbolics and their related compounds, the cresols, have been popular for many years because they are good at killing most bacteria and fungi, including spores. They are less effective against some viruses. They also have the advantage that they are less likely than, say, the hypochlorites, to be inactivated by residual dirt and organic matter, so they are particularly useful on rough walls and floors that cannot be got spotlessly clean. Because of their toxicity both to skin and via the respiratory tract, they must be used with extreme care and their smell lasts for days. This is in itself no bad thing as the house has to be rested anyway.

The liquid chemical disinfectants that I have crudely labelled 'all the rest' include the halogens, e.g. sodium hypochlorite, iodine and the iodophores, and the quaternary ammonium compounds. They are less toxic than the phenols and cresols and extremely effective against bacteria, viruses and fungi provided that the surfaces are completely clean, since they are readily inactivated by the presence of organic material. They are thus suitable for use in a smooth-surfaced calf house after power-hosing or steam-cleaning although in the latter case further disinfection would hardly be necessary.

If it is possible to seal up the building entirely it probably pays to achieve almost complete disinfection by fumigating the building with formaldehyde vapour. This is best done by heating solid paraformaldehyde since the heat can be switched on from outside the building. The method is described in detail in MAFF Leaflet 645 (1981) but I would recommend novices to seek advice on the spot. The method is clearly inappropriate for naturally ventilated calf houses or simple shelters.

Fumigation with formaldehyde is not particularly effective against the spores of the fungus that causes ringworm. Cresols are a little better, but ringworm spores have an uncanny ability to survive in cracks in walls or in woodwork. A flame gun will kill spores in walls. The best way to kill ringworm spores in pens is to clean them and leave them out in the sunshine.

Rest

It is difficult to recommend an ideal rest period for a calf house after cleaning and disinfection. From a strictly health point of view, the longer the better, but that is no help to the farmer trying to earn a living. I would say that a period of two weeks would be the very minimum. After an outbreak of infection with a particularly severe organism, such as *Salmonella dublin* or *typhimurium*, I would recommend that the period be increased to four weeks. The more sunlight and fresh air that get into the building during this time the better.

Further reading

Bruce, J.M. (1975) 'Natural ventilation of cattle buildings by thermal buoyancy' Scottish Farm Buildings Investigation Unit. 42

Ministry of Agriculture, Fisheries & Food, (1981) Leaflet 645. *The Cleansing and Disinfection of Calf Houses.* (1982) Leaflet 836. *Natural Ventilation of Cattle Buildings.*

Mitchell, C.D. (1978) *Calf Housing Handbook.* Scottish Farm Buildings Investigation Unit.

Sainsbury, D. and Sainsbury, P. (1979) *Livestock Health and Housing* Bailliere, Tindall, London.

6 Common diseases and their recognition

The good farm veterinary surgeon is in a position similar to that of the cobbler who sang 'the better my work, the less I earn,' for his aim is to help the farmer control disease at least as much by advice and forward planning as by the sale of vaccines and other preventive medicines.

One essential of disease control is that the stockman should not only recognise disease in its earliest stages but also be able to make a reasonable preliminary diagnosis and prognosis (i.e. presume what is wrong and what will happen next respectively). He can then decide first whether (a) to do nothing, (b) to treat the animal himself, or (c) to call his veterinary surgeon and then what steps he should take to stop the disease spreading to the other animals.

This chapter is designed more for the stockman than the veterinary student since it concentrates on the clinical signs of the most important diseases seen, especially in housed calves, and on first aid and nursing in sickness and convalescence. It does not attempt a comprehensive catalogue of calf diseases and it does not follow the pursuit of a diagnosis through post mortem examination, microbiology and clinical pathology. For those seeking a comprehensive study of calf disease I can recommend a recent translation into English of the excellently written and illustrated book by Ludwig Schrag and others, *Healthy Calves, Healthy Cattle* (1982).

Signs of health and disease

A lot can be learnt about the general health of a group of calves by studying the animals from outside the pen. The healthy calf has an alert appearance, bright eyes and a shiny coat. Its ears are pricked and it shows a keen interest in its surroundings (Fig. 20(a)). If it has been lying down when the stockman arrives, say, first thing in the morning, it will usually rise, arch its back and stretch its limbs in a generally contented manner. The sick calf will be dull and unresponsive to the stockman or other calves. It may stand or lie apart from the rest of the group. Its coat may lack lustre and there may be scouring or discharge from the nose and eyes. The calf with abdominal pain adopts a typical 'tucked in' appearance (Fig. 20(b)) and is reluctant to move. It may also grind its teeth. The calf with pneumonia or some other severe respiratory disease will show clear signs of difficulty in breathing (dyspnoea). It may stand with its forelegs splayed out and its head and neck extended, literally fighting for breath. If it is lying down, again the neck will be extended and the nostrils flared and, in extreme cases, it may have its mouth open and its tongue out (Fig. 20(c)).

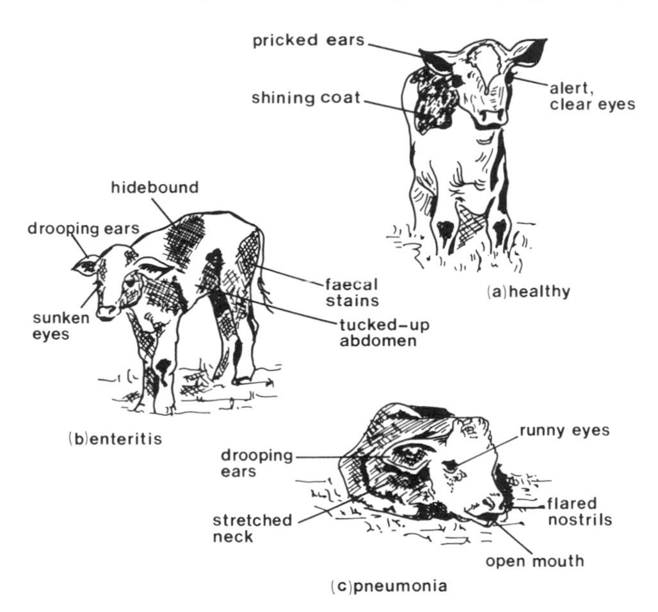

Fig. 20. Signs of health and disease in the calf. (*a*) Healthy: bright eyes, pricked ears, alert appearance, shining coat, skin freely movable. (*b*) Enteritis: 'tucked up' abdomen, faecal staining and hair loss around tail, drooping ears, sunken eyes, dull coat, hidebound. (*c*) Pneumonia: runny eyes and nose, laboured breathing, neck stretched out, flared nostrils and open mouth.

If calves are being fed milk replacer by bucket the stockman will of course notice as soon as the calf goes off its food and this is one of the earliest signs of incipient disease. If the calves are reared in groups he should first take time to observe whether any calf is unnaturally quiet or reluctant to feed, and then get in among the animals. Healthy calves will either approach him inquisitively or run round the pen more in excitement than panic, often kicking both hind feet into the air. The sick calf will be indifferent to all this activity, even if it gets stepped on in the rush. The calf that is stiff or lame will also reveal itself at this time.

Table 15 Some common calf diseases presenting generalised signs of ill-health

Clinical sign	Septicaemia	Infectious enteritis	Nutritional scours	Salmonellosis	Coccidiosis	Lead poisoning	Furazolidone poisoning	Enzootic pneumonia	I. B. R.	Calf diphtheria	Meningitis	Vit. E/selenium deficiency	Cerebrocortical necrosis
Body temperature, elevated	C	S		C			C	C	C	C	C		
General appearance,													
loss of appetite	C	S	S	C	S			S	C	C	C		S
weak or collapsed	C	S	S	C		C	C	S	S			C	C
nervous				S		C	C				C	S	C
Posture, tucked-up abdomen	S	C	S	C		C					S		
spread forelegs								S	S				
splayed shoulders												C	
extended neck & head								C	S				
Eyes, sunken		C	S	C	C								
lachrymating								C	C				
mucus and pus				S				S	C				
Nose & lips, inflammation				S			C	S	C	C			
pus				S				S	C	C			
Faeces, diarrhoea	S	C	C	C	C						S		
mucus and/or blood		S	S	C	C		C						
Breathing, laboured	S			S		S	S	C	C			C	
Hair coat, dull, rough		S	S	C				S	S				
bald patches round anus		S	S	S									
Skin, 'hidebound'	C	C	S	C	C			S					
Navel, hot and swollen	C										C		
Joints, hot and swollen	C										S		

C = commonly seen, S = sometimes seen

If the stockman has any suspicion that a calf may be less than entirely fit he should give it a thorough, systematic clinical examination. It is dangerous to leap to a diagnosis from the first, most obvious clinical signs. Table 15 lists in columns the most common calf diseases that present generalised signs of ill-health. Reading down the rows you have a check list of those signs that should be looked for in all cases of general sickness. If the junction of each column and row is marked with a C it means that the clinical sign is commonly associated with that disease, if with an S the particular sign is sometimes observed, if unmarked the clinical sign is seldom if ever associated with that particular disease.

The first thing to do is to take the animal's temperature, if only because the thermometer should be in the rectum for at least 90 seconds and you can carry on with the examination while you are waiting for it to warm up. Normal body temperature is between 38.5 to 39.0°C (101.3 to 102.2°F). Any temperature over 39.5°C (103°F) should be considered as indicative of disease, probably infectious disease. If rectal temperature is below 38°C (100°F) the calf may be in a state of shock and dehydration. If you do not have a clinical thermometer at the time it is useful to feel the temperature at the base of the ears. In a fevered calf these may feel unaturally hot (although this is not invariably the case). The ears of a shocked, hypothermic calf feel unnaturally cold.

Next observe the eyes. Are the mucous membranes around the eye reddened and inflamed? If so, is this present in one eye or both? If only one, it suggests irritation with a foreign body, usually from the straw, or possibly the early stages of New Forest Disease. Inflammation in both eyes suggests a generalised (systemic) infection. Very pale mucous membranes suggest anaemia. Is there a discharge from the eyes? Is the discharge watery (lachrymation), or purulent (pussy)? Are the eyes sunken in their sockets? If they are, the calf is dehydrated.

Now observe the nose and mouth. Is there any inflammation or ulceration around the nose, lips or gums? Slight inflammation around the nose is quite common in young calves, associated with a large number of disease states both mild and severe. Is there a discharge from the nose and is it watery, mucous or purulent?

Now turn your attention to the body. Pick up a fold of skin. Does it move freely and spring back quickly into place when you release it or does it feel as though it is 'stuck' to the underlying muscle? If so, the calf is suffering from dehydration and is said, rather aptly, to be 'hidebound'. At this time, look for evidence of parasitic infestation of the skin and hair. Ringworm should be obvious, as should lice or mange in the later stages but early signs such as patchy hair loss, especially around the neck and shoulders, may be overlooked.

Watch and listen to the animal's breathing. If you have a stethoscope, so much the better, but you can learn quite a lot by putting your ear directly against the calf (i) under the trachea or windpipe in the neck and (ii) just forward (at the neck end) of the shoulder blade (scapula). Learn to recognise the normal respiration sounds of a healthy calf and then by experience learn to recognise the various sounds, e.g. bubbles, squeaks and rasps typical of different forms and stages of tracheitis and bronchopneumonia.

Observe the area around the anus and the faeces, paying attention to colour and consistency and look for signs of mucus or blood. If there is blood in the faeces, is it dark or bright red? Is there loss of hair around the muzzle, anus or lower limbs? This may be suggestive of a nutritional disorder leading to scouring.

In any calf less than twelve weeks of age, check the area around the navel. Is it hot and swollen? If so, the calf has acute navel-ill which may lead to septicaemia. Can you still feel a thickened, chronically infected umbilical cord? This may not need treatment but it will need watching. Alternatively, is there a soft, movable, non-painful swelling and does it project out of an 'O' shaped hole in the abdominal muscle? This is an umbilical hernia. If you have any suspicion of navel-ill, or if the calf has shown signs of stiffness or lameness, check the joints of the legs. Are they hot and swollen? Does it appear to hurt the calf if you flex the joint? If so, you suspect that bacteria which entered through the navel have got into the blood stream causing septicaemia which is now localised in the joints.

If the calf has shown signs of lameness, check its feet. Foot complaints are not common in young calves but you should make sure. If the limbs are abnormally set, either too straight or with the fetlocks knuckled over, this may indicate some congenital malformation.

When you have completed the clinical examination you should be able to make a preliminary diagnosis, perhaps with one or two alternative possibilities.

Diseases presenting generalised signs of ill health

SEPTICAEMIA

When an animal is exposed to infection it normally succeeds in restricting the infectious organism close to the point of entry, e.g. the skin or the gut wall and the regional lymph nodes. Septicaemia occurs when the infecting organisms break through the first line of defence and invade the general blood circulation. Once this has occurred, damage can occur almost anywhere and secondary foci of inflammation can break out in, for example, the liver, kidneys, joints or meninges of the brain. The newborn calf is particularly prone to septicaemia by two routes, the immature gut wall and the navel. As indicated earlier, bacteria can pass intact across the gut wall of the newborn calf. The organisms most commonly isolated from calves with neonatal septicaemia of enteric origin are various sero-types of *E.coli*. Calves are almost certain to ingest *E.coli* in the first days of life, although it is largely a matter of luck whether they are exposed to virulent or non-virulent sero-types. In either case, they should be safe, provided that they receive colostrum containing appropriate antibodies.

The calf that does get *E.coli* septicaemia within the first 3–4 days of life is either found dead or in a state of collapse. There may be early signs of diarrhoea but the course of the disease is usually too rapid for that. The prognosis is very poor. Calves can be treated with appropriate antibiotics and electrolyte therapy (see below) but usually the damage is already so great that death is merely postponed. *E.coli* septicaemia in the newborn calf should not, however, be confused with enteritis or enterotoxaemia associated with *E.coli* in the older animal. This is much more responsive to treatment, largely because the animal's own defences are much improved by this time. Prevention of *E.coli* septicaemia depends upon ensuring proper colostrum feeding, rigorous attention to hygiene in the calf house and more especially in the calving pens. There is a mixed polyvalent vaccine and antiserum recommended for the treatment of neonatal calves or even the pregnant cow, so as to transfer passive immunity. Whether or not this can prevent disease in individual cases depends upon whether it protects against the particular sero-type responsible for the septicaemia. If all else fails, it may be necessary to give antibiotics at birth.

If the navel is infected, typically with organisms like *corynebacteria*, *streptococci* and *staphylococci*, the infection may remain localised as 'Navel-ill'. If septicaemia ensues the calves will show reduced appetite, listlessness and an elevated temperature, although they will generally be not nearly so ill as calves with *E.coli* septicaemia. The disease may not in fact be noticed until the infection becomes localised in one or more joints. 'Joint-ill'. At this time the calf is very reluctant to move and the affected joints are hot, swollen and painful. Early treatment with appropriate antibiotics* can often bring about a complete cure. If the inflammation becomes chronic the joints will suffer irreparable and perhaps incapacitating damage.

Prevention of navel-ill/septicaemia/joint-ill is achieved by proper attention to navel hygiene at birth. Treatment of the navel with a solution such as iodine (20% solution) has both a drying (and therefore sealing) and an antiseptic effect. Modern proprietary aerosol sprays work in the same way and have the added advantage that the stockman can avoid touching the navel with contaminated hands.

*Note my use of the phrase 'appropriate antibiotic'. Here and elsewhere I have deliberately refrained from making specific recommendations as to the antibiotic of choice for particular infectious diseases because this is the responsibility of the practising veterinary surgeon. One of the most dangerous beliefs held by some stockmen is that some antibiotics are 'stronger' than others. This is a myth. Antibiotics work by destroying or inhibiting specific bacteria. In other words, if the specific organism is sensitive to the antibiotic of choice, it works. If it isn't, it won't. The veterinary surgeon will usually know in advance which antibiotic is appropriate to a particular disease, but in some cases, especially those involving the many and various sero-types of *E.coli* or *Salmonella*, it may be necessary to run a test for sensitivity.

ENTERITIS AND ENTEROTOXAEMIA

Enteritis is an infection of the wall of the intestinal tract only; enterotoxaemia is an infection of the wall of the intestinal tract by organisms that release toxins (poisons) into the general circulation. In uncomplicated cases of enteritis the first clinical sign may only be diarrhoea. The calf may remain generally bright and continue to eat and drink normally. In more severe cases the diarrhoea may be accompanied by severe abdominal pain (Fig. 20(b)). In cases of enterotoxaemia, calves obviously feel ill, varying in degree from listlessness to collapse.

The natural history of most infectious diseases is for the animal to proceed to recovery, with or without treatment against the specific pathogen, unless it dies of the symptoms before it can effect a cure. Calf diarrhoea is a classic case in point and it is every bit as important to treat the symptoms as to try to destroy the infectious organism.

Dehydration

The most important consequence of diarrhoea is dehydration. To illustrate just *how* important this is, I quote a few more simple figures. The normal healthy calf loses by way of faeces, urine, sweat etc. about 10% of its body water per day. In case of severe diarrhoea it may lose in the faeces every day an amount of water equivalent to 25% of its total body water content. Losses of sodium and potassium, responsible respectively for maintenance of blood volume and normal cellular function (amongst other things), may be even more severe. Even if it is getting normal amounts of liquid to drink, the scouring calf will rapidly become dehydrated and if the scouring persists it will die of shock, which is shorthand for increasing weakness, collapse and finally circulatory failure due, in this case, to loss of blood volume. The signs of dehydration and shock are as follows:

4% weight loss: drying of skin, hair and mucous membranes, concentrated dark yellow urine, slightly sunken eyes.

6% weight loss: extreme drying of skin, hair, mucous membranes, hidebound coat and deeply sunken eyes, calf weak and listless.

10% weight loss: as above, plus signs of shock (circulatory failure), collapse, cold extremities, feeble pulse. Nervous signs, twitching, eye rolling.

Symptomatic treatment of dehydrated calves with fluid replacement therapy can be extremely effective, provided the fluids are given in sufficient quantity. A 50 kg calf that loses 10% of its body weight through acute scours needs, by definition, 5 litres of fluid to replace that lost. The most practical way to replace body fluids is through the mouth, provided that what goes into the stomach doesn't further aggravate the enteritis. There are on the market a number of proprietary brands of combined electrolyte/glucose and glycine powders which can be mixed with water and given to calves to replace lost water and to provide electrolytes (sodium and potassium carbonates and phosphates) and a small source of energy that does not require digestion and is easily absorbed. In the early stages of diarrhoea, calves should be allowed to drink these electrolyte solutions rather than milk in amounts of 1.5 to 2 litres per feed according to the size of the calf and the degree of enteritis. If the calves are too weak or apathetic to drink, about 1.5 litres can be administered by stomach tube. Proper application of the stomach tube takes a little practice. It is important to ensure always that it is in the right place (check that the calf is not breathing in and out through the tube!).

The proprietary solutions have the merit of containing all the major essential elements that need to be replaced in the scouring calf. As a cheap alternative or as a first aid remedy you can make up a solution containing 22 grams of glucose and 4.5 grams of sodium chloride (common salt) per litre. In more homely terms, this amounts to one heaped tablespoonful of glucose and one level teaspoonful of salt in 3 pints of water, fed of course at blood heat. The amount of fluid given depends on the degree of dehydration but 5 litres/day given over three feeds or stomach tubing sessions would not be excessive. In very severe cases the veterinary surgeon may elect to transfuse calves intravenously, but when animals have gone this far their chances of recovery are slim, usually because the dehydration is complicated by an enterotoxaemia.

Bacterial and viral enteritis

The organism most commonly isolated from cases of enteritis in young calves is *E.coli*. It is seldom the prime cause of the disease and is more usefully thought of as an opportunist pathogen colonising the intestinal wall following original damage, probably by a virus (rotavirus or coronavirus) or possibly a food antigen. However, it is fair to say that it is *E.coli* that does the most harm, especially if it is strain that causes a severe enterotoxaemia. In the early stages of acute viral enteritis the calf usually has an elevated temperature, abdominal pain and strains to pass very watery faeces containing undigested coagulated milk constituents (a white/yellow scour). Treatment with antibiotics or sulpha drugs has little effect on the primary viral infection, although it may control secondary, opportunist bacteria. The most effective treatment is to substitute electrolyte/glucose solutions for milk replacer for at least two days, in amounts proportional to the severity of the diarrhoea. There is, of course, a limit to the length of time that a calf can be kept alive on these electrolyte solutions. If it is making a fair recovery, it can be given a 'half and half' mix of the milk replacer (made up to the proper concentration) and electrolyte/glucose solution for the third and perhaps the fourth day.

Coccidiosis

Infectious enteritis involving *E.coli* or *Salmonella* usually occurs within the first two weeks of life. Another possible cause of infectious diarrhoea is coccidiosis. This disease is caused by a single-cell parasite (*Eimeria* species) and infection is usually acquired from contaminated fodder or bedding. The condition is usually seen in older calves (more than 10 weeks of age) and is confined to the intestine so that in the early stages the calves are bright and well but are seen to be constantly straining and passing loose faeces, often containing fresh blood. As the bloody diarrhoea becomes chronic, the calves rapidly lose condition. In severe cases, death from dehydration may occur within a week. Other cases drag on. The faeces are less likely to be bloody but they continue to be thin, watery and particularly foul smelling. Calves with chronic coccidiosis fail to thrive.

Diagnosis is confirmed by examining faeces samples for the presence of the parasite (the coccidial oöcysts). There are several effective drugs to treat the condition. Husbandry measures involved in prevention and control of the disease include moving the calves away from the source of infection, e.g. the bedding or an infected pasture, and ensuring that all drinking bowls or troughs are free from faecal contamination.

Nutritional scours

Improper feeding of young calves can precipitate infectious enteritis in ways discussed in Chapter 3. However, many cases of diarrhoea can simply be attributed to indigestion and the consequences of undigested material meeting the normal microbial flora of the small intestine. The cause may be a change of food, or a change in feeding times or some other upset to routine. In calves on *ad lib.* systems it may simply be due to overconsumption. Calves usually appear well in the early stages and rectal temperature is normal. Faeces are wet, bubbly and yellow to dark brown. Mild cases may require no treatment other than strict attention to feeding routines and hygiene. In more severe cases, calves should be taken off milk and fed an electrolyte/glucose solution for about two days. On the third day they should be given half milk powder at the proper concentration and half electrolyte solution; on the fourth day they should be back to normal.

Poisoning

A rare cause of bloody diarrhoea in calves is poisoning with furazolidone, a popular drug for the treatment and prevention of bacterial enteritis. In my experience, cases of furazolidone poisoning have occurred following mistakes made when incorporating furazolidone into medicated feedstuffs or when the stockman has elected to treat with furazolidone calves that are already receiving it in their milk powder. The drug remains predominantly in the intestine but a small quantity passes into the general circulation, where it inhibits blood coagulation and affects the nervous system. The clinical signs include extreme excitability, twitching, bellowing and exaggerated response to mild stimuli, blood in the faeces and haemorrhages around the mouth and nostrils. If the disease occurs, the source of furazolidone must be removed at once. The prognosis for severely affected animals is not good. It may pay to give all animals in the group a multivitamin injection.

SALMONELLOSIS

Salmonellosis is the most severe of the common forms of septicaemia in young calves. There are very many different *Salmonella* organisms with different degrees of virulence and different sensitivities to antibiotics and other antibacterial drugs (e.g. sulpha drugs, furazolidone). The two most common causative organisms are *Salmonella typhimurium* and *Salmonella dublin*. The latter only causes infections in cattle, the former can and does cause severe disease in a wide range of mammals including man. Salmonellosis in calves constitutes a real threat to the stockman and to the stockman's family.

Two factors make a major contribution to the spread of salmonellosis. Firstly, the organism is itself very resistant and can retain its infectivity in dried-out faeces or on pastures for several months. Secondly, the organism, especially *S.dublin*, may persist for long periods in so-called 'carrier' individuals who are apparently healthy but who may excrete it from time to time in the faeces, particularly under conditions of stress such as calving.

It is curious that some individuals, even very young calves, can become infected without showing clinical signs of disease, whereas in others the disease will progress to death within a few days or less. Nevertheless, the source of an infection with *Salmonella dublin* may be carrier cows or bought-in calves. *Salmonella typhimurium* may come from bought-in calves or from an infected stockman. It must be pointed out, however, that most recovered individuals do not become chronic carriers. Most calves cease to excrete *Salmonella* after about six weeks of age.

A calf is at greatest risk of picking up salmonellosis en route through markets and dealers. First of all, it stands a good chance of coming into direct contact with an infected animal. Secondly, lorries and premises do become infected with these very tough organisms despite more or less effective attempts at hygiene. In these days of buying in calves from market or through a dealer, the probability of buying *Salmonella* is high and it pays to act accordingly. All our bought-in calves are treated for the first three days after arrival with a sulpha drug effective against most (but not all) strains of *Salmonella*. Their first two feeds are of electrolyte/glucose solution and the next two 'half and half' feeds are of electrolyte/glucose solution and milk. Protective clothing is worn by the stockman whenever he is in contact with the calves for the first few weeks of life. He cleans and scrubs his boots in disinfectant on entering and leaving each calf unit and he washes his hands thoroughly after work. I would add that some years ago our present stockman did contract salmonellosis from working with calves, so he knows what it is like! If none of your bought-in calves shows signs of salmonellosis within the first two weeks after arrival you can start to uncross your fingers, although the risk remains up to about six weeks of age.

In the early stages of the disease the calf is dull and has an elevated temperature. There is abdominal pain and diarrhoea which is initially difficult to distinguish from viral or *E. coli* enteritis. Later on, however, the faeces become grey/green or dark brown, usually contain large amounts of mucus and often show streaks of fresh, red blood. Initially the calf will continue to drink, because it is thirsty, but it deteriorates fast. Toxaemia is almost inevitable and septicaemia is common, with all that that implies. Many calves with salmonellosis also show signs of respiratory distress; runny eyes and nose, coughing, laboured breathing and often the condition proceeds directly to pneumonia. There is also some evidence that calves that suffer salmonellosis in the first few weeks of life are more prone to contract severe enzoötic pneumonia 4–8 weeks later.

Treatment involves the appropriate use of an antibacterial substance, based on a sensitivity test, fluid replacement therapy but restricted access to food, and general nursing. In the shocked, dehydrated animal this involves the provision of warmth and comfort, e.g. a clean bed of deep straw, a rug or an infra-red lamp.

There are vaccines which are more or less effective against *Salmonella dublin* and *typhimurium*. When administered to animals that are healthy and not otherwise exposed to stress, these vaccines can confer protection within about seven days, although the calves may show some initial adverse reaction. If such a vaccine is given to a stressed calf that has just arrived from market and is already incubating the disease, it is likely to do more harm than good and may even kill the animal. As a calf-rearer I would be prepared to pay a premium for calves vaccinated a week before they left their farm of origin. I do not, however, vaccinate my animals on arrival.

In Britain, salmonellosis is a reportable disease under the Zoonoses Order (1975) which covers diseases communicable from animals to man. If a farmer suspects salmonellosis, he should contact his veterinary surgeon. It is then the responsibility of the veterinary surgeon first to confirm the diagnosis and then to report the disease if the diagnosis is positive. In normal circumstances no restrictions are imposed on the farm.

Following an outbreak, recovered calves and all contacts should be isolated, ideally until they are about 8 weeks of age; when the calves move out, the unit should, of course, be thoroughly cleaned, disinfected and rested as described in the previous chapter.

ENZOÖTIC PNEUMONIA

The word enzoötic means only that the disease is always with us. It does not come and go in the form of isolated epidemics. Enzoötic pneumonia cannot be attributed to a simple organism or small group of organisms. This means that, in the light of present knowledge, it is not possible to control the disease by a process of eradication or vaccination either on the individual farm or on a national basis. Control of the incidence and particularly the severity of enzoötic pneumonia is largely a matter of good husbandry. The mortality rate from uncomplicated cases is quite low but growth and food conversion efficiency are always seriously interrupted and sometimes the animal is stunted for life.

The primary cause of the disease is infection with one or more of a number of viruses or other small organisms called *Mycoplasma*. It is fair to say that every young calf is likely to experience one or more infections of this type in much the same way as every child is likely to catch colds when it first goes to school. When calves are reared in small groups, extensively and in fresh air, these primary infective agents do little more than cause a brief 'chesty cough' similar to that experienced by otherwise healthy schoolchildren. When the environment is bad and as a result the magnitude of the challenge is too high or the resistance of the animal is too low (Chapter 4), then the severity of the initial disease is increased. Primary damage to the lungs and bronchioles attracts opportunist bacteria like *Pasteurella haemolytica* which thrive on damaged tissue and in areas where the oxygenation of the lungs has been reduced, and the disease goes from bad to worse.

Clinical signs of the disease in the early stages include coughing, and an aqueous discharge from the eyes and nose. A purulent discharge from the nose is evidence of secondary bacterial infection. Body temperature is elevated (about 41°C). Some calves recover spontaneously from this initial mild attack within 2–3 days. Other calves appear to be on the road to recovery and then suffer a relapse associated with secondary bacterial invasion of the damaged lung tissue. At this stage, body temperature is again elevated to about 41°C, the calves lose their appetite and become disinterested in everything but the struggle for breath. The clinical signs of dyspnoea (laboured breathing) have been described earlier in this chapter (Table 15 and Fig. 20(c)).

In home-reared calves enzoötic pneumonia classically appears in the period after weaning (6–10 weeks of age) when calves are mixed for the first time. Reasons for this were discussed in Chapter 4. In bought-in calves pneumonia often appears to develop earlier (3–6 weeks), perhaps because infection is acquired earlier, or because passive immunity declines earlier to an inadequate level or perhaps because bought-in calves are, in fact, older than they look.

Treatment of affected animals is most effective when carried out early. This presents something of a dilemma because in good surroundings most healthy calves will recover spontaneously from the primary condition. Broad spectrum antibiotics do not kill viruses, although tetracyclines, for example, are effective against mycoplasma. It is expensive to use antibiotics when they are not needed and also bad practice, since it might encourage the development of antibiotic-resistant strains of bacteria. As a general rule I would not advise the use of antibiotics the first day a calf is seen to have an intermittent cough, a slight clear discharge from the nose and eyes and a temperature no greater than about 40.5°C, particularly if the unit has a good record in terms of spontaneous cures of enzoötic pneumonia. If the temperature exceeds 41°C, or does not fall after 24 hours, or if the calf loses its appetite and becomes severely depressed, then it is essential to begin a course of treatment with broad-spectrum antibiotic lasting at least three days, even if the calf appears to recover sooner. Your veterinary surgeon will advise on this.

Prevention of enzoötic pneumonia by attention to husbandry methods and building design has been discussed at length in Chapters 4 and 5. Vaccines are available but the general impression at present is that they are of little value in most cases of enzoötic calf pneumonia, at least in the U.K. This remark should not be taken as a condemnation of the vaccinal approach. The disease undoubtedly requires the presence of a primary pathogen and if effective vaccines could be produced to confer protection against all or most of the primary pathogens then the disease would cease to be of importance. In a welfare context however, it is perhaps worth stating that the absence of effective vaccines forces the farmer to keep his calves in a reasonably healthy environment. Improved vaccines might make survival possible in the worst of slums.

When the incidence and severity of enzoötic pneumonia are very high it is clearly necessary to rethink the whole husbandry system. This may, however, not be possible until the end of the winter. In these circumstances it will probably be necessary to give all the animals in the group some antibiotic cover, either in the food or by injection throughout the time that some calves are actually showing signs of acute disease.

INFECTIOUS BOVINE RHINOTRACHEITIS (IBR)

IBR is an acute infection of the nose, eyes, throat and trachea of cattle. It was first recognised as a disease of cattle on feedlots in North America by the popular name of 'red nose'. It is however becoming increasingly common (or at least more frequently recognised) in young housed calves in the U.K. and continental Europe. The primary pathogen is a herpes virus. The incidence is high and the virulence of the disease appears to increase with re-infection. In the early stages calves show a high temperature (41–42°C) and general signs of malaise. There is usually a profuse aqueous discharge from the nose and eyes which are extremely inflamed (hence 'red nose'). The calves may show difficulty in breathing and very harsh sounds can be heard from the trachea. Uncomplicated cases recover in a few days, usually helped by antibiotics to guard against secondary bacterial infection. However, if this does occur and the disease progresses to pneumonia, the prognosis becomes poor. IBR can occur simultaneously with, or perhaps precipitate, salmonellosis in very young calves. Such animals have severe damage both to the gut and respiratory tract and are very ill indeed.

The number of farms affected by IBR in the U.K. is still quite small but on those farms the condition can cause serious losses. Fortunately, the disease is associated with a single specific organism and there is a live, intranasal vaccine. This vaccine appears to have a good protective action and can be used on apparently healthy animals even if they are incubating the disease, provided that they also receive antibiotic cover.

CALF DIPHTHERIA

This relatively uncommon condition usually affects calves during the first few weeks of life. It is caused by an organism *Fusiformis necrophorus* found in heavily contaminated bedding or dirty pens (though the disease can occur in the best managed units). The infection probably gets in through a mild abrasion in the mouth, so only one or two animals tend to be infected. Initially there is severe inflammation of the mucous membranes of the cheeks, lips and throat. This proceeds to ulceration and death of the tissues which become greyish/yellow in colour and foul smelling. Because the condition is obviously very painful, the calf will have difficulty in swallowing, it will dribble and be reluctant to eat. If treated quickly with antibiotics the calf will usually recover without showing generalised signs of ill-health. It is however important to be able to differentially diagnose this condition from enzoötic pneumonia and IBR or an infection resulting from a foreign body trapped in the mouth.

LEAD POISONING

Practically everybody who rears calves knows that they can die of lead poisoning, yet cases continue to occur. The most obvious sources are lead-based paints. It is most unlikely that anyone would newly decorate a calf house with a lead-based paint but quite often, as old paintwork peels, lead-based products come to the surface. Often the source is not immediately apparent but turns out to be something like an old car battery or the remnants of a leaded window. Acute lead poisoning can present as sudden death. In other cases the first sign is when the calf throws a fit. It becomes manically excited. It appears to be blind and runs into walls or even attempts to climb them. It may press its head against the wall and be completely oblivious to other stimuli. I assume that an animal which compulsively presses its head against the wall has an agonising headache. Calves that reach this stage usually collapse, go into muscular spasm and die quite quickly.

In the less acute form calves may be dull and listless, with signs of abdominal pain, tucked-up abdomen, grunting and grinding the teeth. Unlike cases of infectious enteritis, the animals are usually constipated. Lead poisoning can, however, be confused with cerebrocortical necrosis (q.v.), hypomagnesaemia, which is rare in young calves except in suckler herds (Chapter 8), meningitis resulting from septicaemia or furazolidone poisoning.

If you suspect lead poisoning then contact your veterinary surgeon. As a first aid remedy you may try drenching the animal with a solution containing 100 g (about 4 oz) magnesium sulphate (Epsom salts).

VITAMIN E/SELENIUM DEFICIENCY (WHITE MUSCLE DISEASE)

The various tocopherols that one generally includes under the title of vitamin E are linked with the element selenium in controlling the metabolism by the body of fatty compounds, particularly unsaturated fatty acids and the phospholipids that make up cell membranes. In the event of a deficiency of vitamin E or selenium, the breakdown and repair of cell membranes gets out of control. This is most apparent in the membranes of skeletal muscle or sometimes heart muscle. It follows therefore that clinical signs of deficiency include muscle weakness and cardiac failure.

The condition is very rare in properly fed, early weaned calves but it can occur in sucklers' calves on selenium-deficient pastures. It can also occur in veal calves or possibly barley beef animals if the compound feed has been stored too long and the potency of the tocopherols has decreased. As mentioned above, the clinical signs include general muscular weakness. The animals lie down a lot, have difficulty in rising and are reluctant to move. Typically, the weakness is most apparent in the muscles of the shoulder. The animals may collapse into a lying position with their front legs splayed out. When they stand, the trunk may sink so that the shoulder blades stand out prominently. The urine is usually very dark due to the presence of pigments from degenerating muscle. Signs of nervous disorders are common. Treatment of early cases by injection of vitamin E and selenium can be spectacularly successful. The effect of the injection is long-lasting so it can also be used for long-term prevention. However, it is necessary to proceed with care, as selenium is extremely toxic in excess.

Calves with vitamin E/selenium deficiency occasionally simply drop down dead with a heart attack, having presented no clinical signs previously. This can happen in bucket-fed calves in the middle of or immediately after a meal. Some calves have been known to drop dead (of excitement?) even before the meal arrived. There are many reasons why a calf may have a heart attack, but if I had a calf drop dead with cardiac failure in this way and I had not previously treated the animals with vitamin E, I would probably treat the rest. If two dropped dead for no other good cause, I would certainly treat the rest.

CEREBROCORTICAL NECROSIS

This is a rather rare condition but one that needs to be considered as a possible cause of weakness or nervous disorder in calves. It appears to be caused by a deficiency of thiamine (vitamin B_1). The story is probably not that simple, but it is certain that early cases do respond to treatment with vitamin B_1. The normally developing rumen should provide enough thiamine for the calf, but abrupt changes in feed may give rise to upsets in rumen fermentation and so reduce thiamine production. Other cases may be due to the presence in mouldy feedstuffs of thiaminases, enzymes which destroy thiamine. The disorders of metabolism that result from this deficiency cause swelling in the cerebral cortex of the brain. Affected calves are dazed and nervous. They show muscular tremors and poor vision. Typically they will bump into things if forced to move about. Occasionally they will show signs of acute mania. In the terminal stages the calves will lie flat out on their sides, making weak paddling movements with their legs. Death may occur within 3–4 days of the onset of symptoms.

As with so many nutritional deficiencies, successful treatment provides the best means of diagnosis, for while the natural history of an infectious disease is to proceed to recovery with or without treatment, the natural history of a deficiency disease is to get progressively worse unless the deficiency is remedied.

Skin conditions

RINGWORM

Ringworm, or infection with the fungus *Trichophyton verrucosum*, is an extremely common condition of young housed calves, particularly in cool, damp winters when the relative humidity is high and the animals have no access to direct sunlight. The spores of the fungus survive for months or years in cracks in walls, or in wood, or in straw and this is probably the initial source of infection. However, the disease is extremely contagious and in group housing, once one calf has got it, the infection is likely to spread to the whole group by direct contact between animals. First signs usually appear around the eyes and elsewhere on the head and on the neck, presumably because these are the areas the animals deliberately and accidentally rub against each other and against the structures in the building. Initially the condition presents itself as small, round, raised nodules on the skin which may be slightly inflamed. Soon the hair falls out leaving round, bald patches. Initially these are pink in colour but a dry, grey crust of dead skin forms over the area quite rapidly. The outside ring may remain active and continue to enlarge while the centre proceeds through the crusty stage and finally starts to grow hair again. In most cases, however, each area seems to grow to a certain size and stop, even though new sites of infection continue to break out.

Eventually calves develop an effective and lasting immunity to ringworm. This may take four or five months and usually seems to coincide with the animals going out to grass. Here the self-cure is partly immunological and partly environmental. The severity of the condition tends to be worse in damp and dismal buildings than out of doors in the fresh air. Treatment can be by local application of fungicides or by incorporating a systemic fungicide in the food. The latter remedy is rather expensive. My graduate students and I have spent a lot of time simply watching calf behaviour (Chapter 7). It is our impression that ringworm is not particularly distressing to calves unless it is very severe. It certainly does not appear to irritate them as much as lice. If the condition is not severe and the calves will soon be going out to grass I do not believe treatment is necessary. I must however point out that the condition can be transmitted to man, especially children. For them, the condition it unsightly, distressing and can lead to permanent hair loss. For that reason alone it is perhaps advisable to keep calves as free from ringworm as possible.

LICE

Infestation with lice is extremely common, indeed almost inevitable in calves housed over winter. Lice are small, blood-sucking insects. The parasites and their eggs (nits) are visible to the naked eye as small, white specks. Lice are usually transmitted direct from one animal to another. They do not survive very long in empty buildings. For this reason it is useful to dust bought-in calves with anti-louse powder on arrival.

First signs of louse infestation include slight hair loss behind the ears and on the neck and shoulders. The animals rub, lick or scratch the affected areas, although not to excess. If the infestation becomes more severe the coat becomes dry and harsh, areas of hair loss spread and the bare skin becomes rough and scaly. At this stage the irritation is intense and calves spend a lot of time licking, scratching and rubbing. Massive infestation with these blood-sucking parasites can cause quite severe anaemia. Even moderate infections stunt the animal's growth.

The environmental conditions that predispose to infestation with lice are overstocking and high relative humidity. The condition can be particularly severe in veal calves raised in groups in straw yards, as these calves get very hot and sweaty and shed their birth coats at about 6–8 weeks of age (Chapter 9). The combination of sweat and matted hair seems to provide a particularly attractive breeding ground for this parasite.

Lice also thrive on the calf that has already been set back by a debilitating disease such as enteritis or pneumonia. This is because the skin and long, dull hair coat of the sick calf again favour development of the parasite.

Treatment of lice either by local application of powder or a liquid insecticide, or by injection of a systemic insecticide is very effective and is an essential part of good husbandry in all but the mildest of cases.

MANGE

Mange is another skin disease caused by parasites. It is less common than lice but it can be more severe and more difficult to treat. It is important therefore to distinguish the two conditions.

Mange mites are much smaller than lice and are not really visible to the naked eye. They live on or in the skin and feed on body fluids and skin squamae. Their eggs are very resistant to treatment by most insecticides so it is necessary to treat animals repeatedly to kill all mites after they have hatched. The development of the eggs takes 10–20 days, so it is necessary to repeat treatments at 10-day intervals.

The initial clinical signs of mange are similar to those of lice, although the condition often starts around the tail and rump, as distinct from lice which usually starts on the neck and shoulders. The animal may of course be infested with both parasites. Usually the irritation caused by mange is more severe relative to the amount of skin affected than that caused by lice. In advanced cases the skin may become raw as a result of the initial infestation and the attentions of the animal itself. If secondary bacterial infection ensues, the animal becomes covered in weeping, purulent scabs. It is altogether a most unpleasant condition.

Treatment of a calf for mange should be based on a specific diagnosis from microscopic examination for parasites in skin scrapings.

Treatment usually involves regular washings and scrubbing with an insecticide, although it can again be treated, at some expense, by an injectable parasiticide. The disease is transmissible to man.

NEW FOREST DISEASE (INFECTIOUS BOVINE KERATOCONJUNCTIVITIS (IBK))

New Forest Disease, or IBK, is a highly infectious, inflammatory disease which initially affects the mucous membranes of the eyelids but can rapidly spread across the whole eye. Traditionally it is thought of as a disease of cattle in their first summer at grass, since the infective organism is transmitted by flies, but it is being observed increasingly commonly in housed calves. The infection may be brought in by flies but it is probably spread within the building by direct contact and on dust particles. It can occur in housed calves at any time of year, although it is probably more common in mild, muggy conditions when flies are about.

The condition is extremely painful and affected animals are severely distressed. Usually, one eye only is involved. At first, the mucous membranes around the eye are inflamed, the eyelids are swollen and the eye waters copiously. If left untreated, the eye becomes very inflamed, there is a purulent discharge, the cornea becomes completely opaque and the animal becomes blind in that eye. In very severe cases the eye may be completely destroyed. Treatment should be started without delay, mainly because the disease is so acutely distressing but also to save the eye. If begun in time, treatment is effective. Recovered animals develop a lasting immunity which suggests that an effective vaccine may not be too far distant.

Digestive disorders in the weaned calf

Infectious diarrhoea is not a serious problem for the weaned calf but it is prone to a number of diseases which can be grouped under the general heading of indigestion and which can usually be attributed to the fact that it is eating more of a particular sort of feed than its immature digestive tract can tolerate.

BLOAT

Bloat describes the excessive accumulation of gas in the rumen. In normal circumstances, controlled contractions of the rumen drive fermentation gases up the oesophagus at intervals of 1 to 2 minutes. Cattle at pasture may suffer from 'frothy bloat' which occurs when agents in the grass cause a layer of foam to form on top of the rumen contents, trapping the fermentation gases in such a way that they cannot be released by the normal belching process. In young calves this condition is not common and bloat is usually of the 'gassy' variety, i.e. the upper part of the rumen is grossly distended by gas, rather than foam, but the calf fails to belch it up. The exact reasons for this are still not clear, but in most cases bloat seems to follow overeating of highly fermentable foods and some other stimulus to very rapid fermentation, such as a long drink after a period without water. In these circumstances we must assume that gas accumulation is so rapid that it initially overwhelms the normal belching mechanism. Once the rumen is grossly distended by gas, the calf's ability to belch actually diminishes rather than increasing.

The clinical signs of bloat are extreme distension of the left flank, which produces a drum-like sound if you tap it. The calf is in great pain and may die quite quickly if untreated. Mild cases of gassy bloat after feeding tend to subside spontaneously, although the animals must be carefully watched to ensure that they don't get worse. If bloat occurs regularly, the feeding regime should be changed. This usually means more forage and less concentrate. Acute, severe cases of bloat require rapid treatment. The classic veterinary approach to this is to puncture the upper part of the rumen through the left flank with a trocar and cannula. In my opinion this is usually unnecessary for calves. Gassy bloat in calves can usually be relieved by inserting a stomach tube through a hole placed in a wooden gag jammed between the molar teeth of the calf. If you do not have a stomach tube and the calf is clearly very ill, you can usually release a lot of the pressure by inserting the longest, largest hypodermic needle you can find into the rumen through the upper part of the left flank. The release of gas may take a little time and the needle may need to be adjusted once or twice to keep the gas coming, but it is a far more sensible approach than resorting to desperate remedies with a penknife.

IMPACTION OF THE RUMEN

This condition occurs when there is a failure to ferment and degrade material entering the rumen to the point where it can be expelled through the reticulo-omasal orifice and into the abomasum. In young calves the condition is seen most typically in the animal that is eating a lot of stemmy hay or straw. The animal has a pot-bellied appearance. Through the left flank you can feel that the rumen is distended with material that feels dry, solid and tightly packed. It feels, in fact, rather like a bag packed full with chopped silage, which is hardly surprising. Rumen impaction is usually a chronic condition. The animal will be unthrifty-looking and usually constipated. Appetite will be poor or worse.

There are various preparations to stimulate the inactive rumen. Control of the condition mainly involves attention to feeding practices after weaning. When calves are kept in groups and access to concentrates is restricted, the stronger or more dominant calves tend to monopolise the concentrates, leaving the weaker calves nothing but forage. These are the animals most prone to suffer rumen impaction. The remedy may lie in (a) increasing the amount of concentrate fed, (b) increasing trough space for concentrate feeding, or (c) reducing group size and matching animals for age and size. Once again, it is essential to ensure that all calves have access to clean water.

ACIDOSIS AND LAMINITIS

Fermentation in the rumen converts carbohydrate to organic acids. In normal circumstances the principal products are acetic, propionic and butyric acids and the rate of their production keeps pace with the rate at which they are absorbed, diluted or buffered by salts like sodium bicarbonate in saliva. In consequence, the contents of the rumen remain only very slightly acidic (pH about 6). However, if the rate of production of acids by fermentation greatly exceeds the rate at which acids can be absorbed, buffered or washed out of the rumen, then the acidity of the rumen will rise (pH will fall). This can occur when the animal overeats rapidly fermentable foods like barley or other low fibre cereals. It is made worse if the calf does not have free access to water and a reasonable supply of palatable forage, preferably unchopped and certainly not chopped shorter than about 5 cm (2 inches) in length. The importance of long forage is that it stimulates rumination and therefore salivation and thus increases the flow into the rumen of water and buffers like sodium bicarbonate.

If, as a result of too rapid fermentation, pH in the rumen drops below about 5.0, things go from bad to worse. Protozoa and bacteria associated with normal stable rumen fermentation are killed and lactobacilli, which thrive in more acid conditions (at lower pH), become dominant. The calf with acute acidosis is clearly distressed. There may be bloat and abdominal pain. Rectal temperature is usually normal but the animal typically shows a marked increase in the rate and depth of respiration as it tries to cope both with the increased production of acids in the rumen and the absorption of lactic acid into the blood stream. Acidosis associated with severe overeating can be fatal. More often the animals recover from the acute attack but then suffer from reduced appetite and indigestion for some days or weeks thereafter before the microbial population of the rumen gets back to normal.

A serious complication of ruminal acidosis is laminitis. When the rumen contents are very acid, lactic acid produced by the microbes (which is a 50:50 mixture of L- and D-lactic acid for those who are interested) and probably other breakdown products are released into the circulation in amounts sufficient to inflame and damage capillaries and small blood vessels, most seriously in the laminae of the hooves – the junction between the horny surface of the hoof and the live, sensitive tissue underneath. In acute laminitis one or more feet (often all four) are extremely painful and hot to the touch. Calves are reluctant to move. They stand stiffly, often with the front legs crossed, and take their weight off the affected foot or feet as far as possible.

Severe or chronic laminitis can lead to separation of the sensitive and horny laminae. This distorts the subsequent growth of the foot and can lead to a number of further complications which are outside the scope of this book (see Weaver, 1981). In any event, growth of the animal is impaired at least temporarily and sometimes permanently. Since laminitis in young cattle only occurs as a result of ruminal acidosis, it can only be controlled by ensuring a controlled rumen fermentation.

Even if the intention is to rear calves predominantly on cereals (e.g. for barley beef) it is essential to provide at least 15% of dry matter intake in the form of long fibre to stimulate normal rumen development, normal rumination and salivation. Some feed compounders incorporate sodium bicarbonate into their calf feeds to provide 50–100 g/day for calves in the first weeks after weaning. This can be effective in reducing the risk of acidosis but it makes more sense to stimulate the calf to dose itself with its own salivary sodium bicarbonate produced from sodium ions which can be recycled and carbon dioxide which is a waste product of tissue oxidation

Diseases contracted at pasture

This book is mainly concerned with rearing calves from the dairy herd from the time they are taken from their mothers until the time that they are put out to pasture for their first summer so it is not appropriate to consider in any detail the wide range of disease conditions that they may experience at pasture. This section will only outline very briefly what can occur and what preventive measures may be adopted before turning out the calves. Chapter 8 discusses in rather more detail some of the conditions that can affect beef calves born to suckler cows at pasture. For a fuller description of conditions that can affect cattle at pasture I would recommend either Schrag (1982) or a standard textbook on Veterinary Medicine such as Blood, Henderson and Radostits (1979).

PARASITIC ROUNDWORMS

The most serious of these is the parasitic lungworm *Dictyocaulus viviparous*. The adult worms live in the trachea and bronchioles and severely interfere with normal respiration. This common condition is known in the UK as husk or hoose. Affected animals lose condition very fast and weight gains made during the winter and spring can be wiped out in a few days. The adult worms lay their eggs in the bronchi and trachea. These hatch into larvae which are coughed up, swallowed and excreted in the faeces. Animals contract the disease by eating larvae from infested pastures. The disease can be treated or controlled by regular dosing with anthelmintic drugs, but the most secure protection is achieved by the use of an oral vaccine (Dictol) produced by irradiating larvae. Two doses are required, separated by an interval of approximately four weeks and calves should be protected from sources of lungworm infection until two weeks after their second dose. In other words, both doses should be given before the calves are turned out to pasture. In areas where husk is endemic, vaccination becomes almost essential.

Parasitic roundworms of the gut can also be extremely debilitating for young cattle. However, if the level of pasture infestation is not severe, cattle can build up a satisfactory immunity to these roundworms such that their effect is small. With proper pasture management it is possible to rear calves at pasture without recourse to drugs against parasitic roundworms of the gut. However, it is, of course, imperative to treat animals before they start to lose condition. A veterinary surgeon can assess the degree of parasitic infestation from a count of egg numbers in a sample of faeces.

CLOSTRIDIAL INFECTIONS

Bacteria which are classed together under the generic name *Clostridium* are anaerobic, which means they thrive in the absence of air, and form spores which means they can survive for many years. These organisms live in the soil. Clostridial diseases of cattle include tetanus and blackquarter or blackleg. Infection may be through small wounds or, in the case of blackquarter or blackleg, invasion from the gut. Sheep are more prone than cattle to these and other clostridial infections presumably because they graze much closer to the ground than cattle and are more likely to ingest soil. Routine vaccination of sheep against clostridial infections is considered a must. Cattle are less likely to be routinely vaccinated, the decision being based usually on local knowledge. Again, this is a matter for discussion between the farmer and his local veterinary surgeon.

MISCELLANEOUS CONDITIONS

New Forest Disease is a common condition of calves in their first summer at pasture (see p. 138). Various deficiency diseases can occur. Hypomagnesaemia is a possibility shortly after turnout. Deficiencies of stored trace elements like copper, cobalt or selenium are likely to occur later in the summer when body reserves fall to critical levels. The prevention and treatment of these conditions are outside the scope of this book but are covered excellently by Underwood (1980).

Further reading

Blood, D.C., Henderson, J.A. and Radostits, O.M. (1979) *Veterinary Medicine* (5th ed.) Bailliere Tindall, London.

Schrag, L. (1982) with H. Erz, H. Messinger, F. Wolf and J. Taxacher. *Healthy Calves, Healthy Cattle*. Verlag L. Schaber. D-8355. Hengersberg.

Underwood, E.J. (1980) *Mineral Nutrition of Livestock* Commonwealth Agriculture Bureaux, Slough.

Weaver, A.D. (1981) *Lameness in Cattle.* 2nd ed. Wright/Scientechnica, Bristol.

7 The development of behaviour

At the beginning of this book I quoted 'Five Freedoms' as a set of standards for satisfactory farm animal welfare. These were, freedom from (1) hunger and malnutrition, (2) thermal and physical discomfort, (3) injury and disease, (4) suppression of 'normal' behaviour and (5) fear and stress. Chapters 2 and 6 dealt with the first three standards. This chapter will consider the development of normal behaviour in calves and the extent to which it is affected by the system of husbandry imposed on the animals.

Behaviour is what an animal does. In precise terms, it is the response of an animal to stimuli which it senses either from its environment or from within its own body. The individually penned calf that moos when the stockman arrives in the morning is responding to his presence and calling for milk. The calf in a yard or field that suddenly kicks its hind legs in the air presumably 'sensed a stimulus from within its own body,' or, avoiding the rather ponderous jargon of the behaviourists, we can say it simply felt like it.

Much of the study of animal behaviour involves observing the animal in such a way that the presence of the observer does not, of itself, influence that behaviour. Television and cine cameras are very useful here. Observing how an animal acts does not, of course, necessarily reveal how it feels. In the example given above, we can safely assume that the actions of a healthy calf just before it gets its morning feed are those of pleasurable excitement. If, however, we observe a veal calf licking the wood in its crate for long periods, can we necessarily assume that the animal is deeply disturbed? It is, after all, no less purposeful an activity than chewing gum. It is therefore relatively easy to study animal behaviour but often extremely difficult to establish motivation.

There are other objective ways of measuring whether and how much a calf has been disturbed by a particular environmental stimulus. The most popular methods involve continuous recording of heart rate or sampling the blood for any increase in release of steroid hormones from the adrenal cortex. Heart rate will, of course, increase in response both to pleasant and fearful sensations. Release of steroid hormones is increased generally by 'stress' which can include fear, pain, cold, exercise or even underfeeding. These so-called 'objective' measurements of an animal's response to environmental stress do need careful and cautious interpretation and, once again, it is essential to ensure that the experimental approach (e.g. taking blood samples) does not, of itself, cause a stress.

It is a safe assumption that the development of behaviour in an animal is determined by its genetics and by its environment; safe, that is, provided one does not extend this eminently logical argument to man and so incur the wrath of those who have political reasons for believing otherwise. One would therefore expect the development of behaviour in a young calf to be determined by its breed, its sex and the system of management imposed on it in the first few weeks of life.

Oral behaviour

The calf uses its mouth not only for the essential tasks of eating and drinking but also as a major instrument for investigating the environment. The mouth is also involved in most sorts of apparently abnormal behaviour, pen licking, cross-sucking, urine-drinking, etc. Let us consider the various things a calf does with its mouth in some detail.

SUCKING

At birth the calf has an instinctive (i.e. unlearned) drive to seek its mother's teat and obtain first colostrum and then milk. It appears to rely in general terms on sight, warmth and probably smell, although it very likely has no preconceived notion of what a teat looks like. The calves of suckler cows of traditional beef breeds with small udders usually find a teat with a little difficulty, apparently by working their way to that part of the abdomen furthest from the ground. The Friesian cow has, in a very few generations, been selected for a large udder with teats far below the abdominal wall. It appears that the evolution of behaviour in the Friesian calf has not yet come to terms with this change in the shape of its mother, and calves born to Friesian cows are notoriously slow to find a teat. This is obviously important because of the crucial necessity for a calf to receive adequate colostrum in the first day of life.

The instinctive sucking behaviour of the new-born calf is, of course, reinforced by learning if it continues to live with its mother. Calves that are weaned onto bucket-feeding immediately after birth lose the innate sucking instinct by about 5 days of age (range 3 to 9 days). In other words, after about five days a calf that has got used to a bucket would not, on being re-united with its mother or another cow, instinctively seek out a teat as being the essential source of food. Learned behaviour has overwhelmed instinctive behaviour within the first week of life. This is not to say, of course, that once calves have been weaned onto a bucket they will no longer drink from a teat. That is obvious nonsense. The point is that calves which are put back onto the teat, either in an artificial rearing system or on a multiple-suckling system with beef cows, have to re-learn from experience (and perhaps from observation of others) what to do.

A calf at pasture with its mother will take 4 to 10 feeds per day varying in quantity from about 0.8 to 2.5 litres at a time, rather depending on the size of the calf and the amount of milk available. Twin calves, or calves in a multiple suckling system, tend to drink more often, presumably because less milk is available at any one meal. Average meal size is about 1.5 litres, which corresponds to abomasum size in the young animal, i.e. most calves take a bellyful if they can. Meals of up to 2.5 litres are usually taken by the bigger, stronger calves but there is no doubt that some calves of suckler cows on good clean pasture can suffer from nutritional scours if their mothers have a lot of milk. There is some physiological evidence that calves do not acquire the ability to regulate appetite 'properly', i.e. in response to metabolic needs, until they reach 12 weeks of age. Prior to this time, their only constraint on food intake is the volume of their stomachs. It is not surprising, therefore, that young calves are prone to indigestion and the consequences thereof, if they are provided with too much highly nutritious food, be it milk or a starter ration.

The fact that the calf elects to take a relatively large amount of milk in relatively few feeds per day (4 to 10) is one reason why it has been possible to develop artificial rearing systems based on twice or even once-daily feeds with milk substitute.

DRINKING

There is little that needs to be said about drinking behaviour. Water consumption depends on various factors: the dry matter content of the food, air temperature and its effect on water loss by evaporation, and the water content of the faeces. Generally speaking, calves regulate their water balance very well, provided they have continuous, or at least regular access to clean water. They are reluctant to drink water from a bowl contaminated by faeces. This can sometimes present problems with newly weaned calves. The water bowl gets contaminated on Saturday night and is not cleaned out until Monday morning. Calves who have not drunk all day Sunday then drop a large volume of water onto a bellyful of dry, unfermented concentrate feed. Fermentation rate suddenly increases and the calves get bloat.

There is some evidence to indicate that Friesian and Holstein calves are easier to wean to a bucket than calves from the beef breeds. Taking a strictly Darwinian view of evolution, I am a little surprised that this has occurred so soon but it is, I suppose, true that there has been selection pressure not only for a strong sucking drive in calves reared by their mothers, but also for a rapid adaptation to bucket-feeding in calves from the dairy herd. In either case, the calves that adapt less well to the system imposed are less likely to survive.

EATING

In the first few days of life calves investigate with their mouths just about every solid object that doesn't run away or fight back. We must assume that they acquire the knowledge of what is and isn't good to eat by experience rather than instinct, aided, in the case of the calf at pasture with its mother, by observation and imitation.

I have already said that the appetite of the young calf for solid food does not appear to be related to its metabolic needs but simply to the amount of space available in its stomachs and the amount of work required to get it in. This point is illustrated by Table 16 which compares the eating and ruminating behaviour of calves getting grass either in the long form, chopped or ground and pelleted.

There was little difference in eating or ruminating behaviour in calves given long or chopped grass; they ate for about 4½ hours and ruminated for 7½ hours. When the grass was chopped and pelleted they ate for only about 2 hours, ruminated about 2½ hours but consumed 80% more food. If we make the reasonable assumption that the taste of the grass was not significantly improved by grinding and pelleting, we reach the very important practical conclusion that the food intake of calves in the first few weeks after weaning is markedly affected by the physical form in which

Table 16 Effect of the physical form of forage on the eating and ruminating behaviour of calves at 6 to 9 weeks of age.
(adapted from Hodgson, 1971).

	Grass pellets	Long or chopped grass
Eating (min/24 h)	132	276
Ruminating (min/24 h)	138	459
Eating (min/kg DM)	85	320
Ruminating (min/kg DM)	70	535
Dry matter intake (g/24 h)	1553	862

the food is presented. If a weaned calf gets most of its food as unprocessed forage (and this includes the weaned calf grazing spring grass) it will grow more slowly than the young calf getting processed food, partly because of differences in the ME content of the different foods (MJ ME/kg DM) but mostly because of differences in DM intake. The optimal balance between long grass or long forage and processed foods depends on circumstances. The dairy heifer calf needs to grow quite fast but she also needs to develop a large rumen and thus a large appetite for forage when she comes into milk. It may pay to restrict dairy heifers to 2 kg of concentrate ration a day and thus stimulate their appetite for forage. There is also evidence that forage intake is improved if the calves are given unrestricted access to clean water before and after weaning.

The farmer who gives his calves unlimited access to concentrate feeds, perhaps because he wants them to calve at two years of age or because he intends to rear them for 'barley beef' at about 12 months of age, runs the risk that his calves will overeat in the first few weeks after weaning and fall prey to the digestive disorders of the weaned calf – bloat, acidosis and laminitis (Chapter 6).

RUMINATION

Rumination describes the process whereby a ruminant animal regurgitates a bolus of food from the rumen, chews on it for about a minute, then swallows it once again. The primary purpose of rumination is to break down the structural elements of the food so as to render it more readily available to the microbes in the rumen. The secondary function is to stimulate the flow of saliva into the rumen bringing water and buffers like sodium bicarbonate, and recycling potential nutrients like urea.

The Brambell Commission (1965) recommended that diets for calves should 'permit of rumination' (para. 145). This rather implies that calves and other ruminants have a 'right to ruminate'. I would consider this an extreme view. Table 16 showed that calves given grass pellets ruminated for 30% of the time of those eating long grass. In my own experience, adult sheep fed entirely on ground, pelleted dried grass ruminated for only about 10 minutes per day, as against 300 minutes per day when they were given unchopped dried grass. It would seem that the major stimulus to rumination is the prolonged presence of long fibre in the rumen, and the length of time they spend ruminating depends on just how much fibre there is rather than any instinctive drive to chew the cud. All ruminants, however, do something called pseudorumination which involves regurgitation movements which bring up liquid, or possibly nothing, a few desultory chewing movements and subsequent swallowing movements if there is anything to swallow. Calves (e.g. veal calves) which get no long fibre will show a few sequences of pseudorumination per day. Whether they are doing something constructive or merely satisfying a basic need, I do not know. In any event, they don't do it for very long.

Calves eat solid food in the first few days of life and begin to ruminate almost immediately thereafter. Given a normal mixture of milk and solid food the calf increases the time it spends ruminating very rapidly. At six weeks of age rumination time may be 70% of that of an adult on the same food; by ten weeks of age rumination time is similar to that of an adult. The effect of rearing systems on rumination will be discussed later. For the moment, we can conclude that the development of the act of rumination is more rapid than the development of functions in the rumen. Presumably the immature rumen is more likely to retain long, undigested fibre and so stimulate rumination.

'PURPOSELESS' ORAL BEHAVIOUR

Use of the mouth for eating, drinking, ruminating, grooming and constructive investigation of the environment may be deemed purposeful oral behaviour. However, calves do a lot of other things with their mouths which would seem to us to be purposeless, such as licking and chewing inedible (or unswallowable) objects, sucking the ears, navels, teats, tails and pizzles of their neighbours and drinking urine (See Fraser, 1980). Cross-sucking and urine drinking are often called 'vices' which rather implies that calves should conform to human standards of good behaviour. Cross-sucking is, however, a potentially dangerous way of spreading infection. Compulsive urine drinkers have, in my experience, tended to show abnormalities of rumen development, so this practice may also be positively harmful.

Calves undoubtedly seek and obtain forms of direct oral satisfaction which have nothing to do with hunger or thirst. A classic example of this is the 'kissing' behaviour of veal calves in adjacent pens (Fig. 21). This occurs almost invariably immediately after the calves have finished a large but short-lived meal of milk from a bucket. Kissing is not a fanciful term; I can think of no other word to describe an activity which involves two animals putting their lips together, sticking their tongues into each others' mouths and sucking. I think it is fair to conclude that veal calves 'kiss' because they are deprived of the prolonged oral satisfaction of sucking. If bucket-fed calves are penned up for about 20 to 30 minutes after each meal, the tendency for this and other forms of cross-sucking is much reduced. It would not, however, be proper to conclude that denying calves the right to suck a teat is an unacceptable denial of the right to display normal behaviour. They may indeed *prefer* kissing. About the only system of husbandry that can be criticised unequivocally

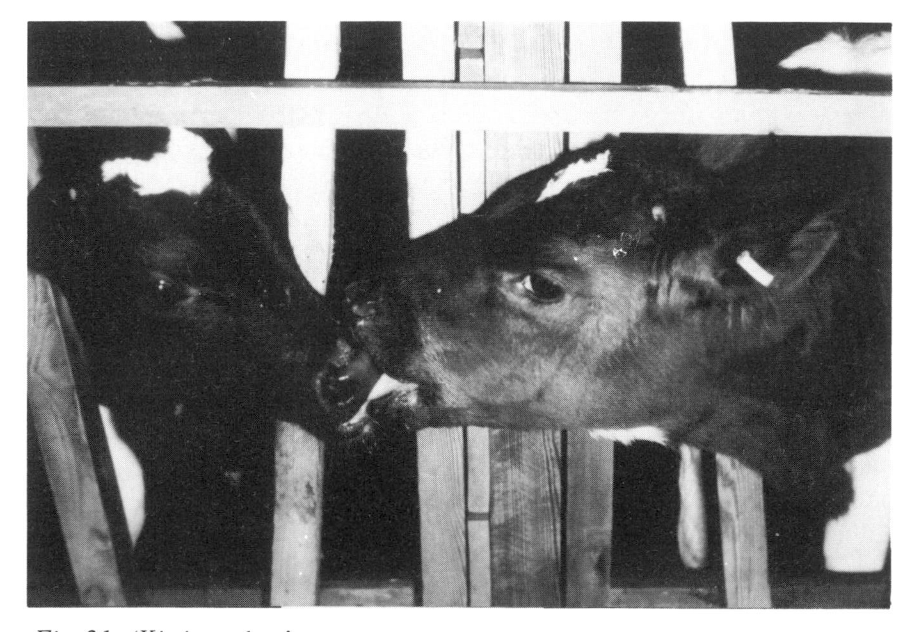

Fig. 21. 'Kissing calves'.

for denying calves oral satisfaction is that which rears veal calves with their heads permanently in plastic muzzles (yes, permanently, they suck up the milk through the holes in the muzzle).

Resting behaviour

The most common lying position for the calf or adult cow is on its brisket with one or both forelegs tucked underneath (Hafez, 1969). Cattle with a functional rumen need to adopt this position if they lie down for any length of time, to keep the gas bubble on the top of the rumen and so avoid getting bloat. The posture taken by the calf when it is lying down also reflects how warm or cold it feels. The calf that feels particularly warm will lie on its side and stretch out its limbs for as long as it can. This form of lying behaviour is seen in calves at pasture on a sunny day and veal calves reared in yards. In the latter case this is partly because veal calves get very warm (p. 80) and partly because, having a less active rumen, they can lie down this way for longer before the rumen becomes uncomfortably distended.

Calves display exactly the same patterns of drowsiness, deep sleep and paradoxical (or rapid-eye-movement, REM) sleep as we do. Young calves sleep for much longer periods than adult cattle. In the wild this was possible because other members of the herd kept watch. During the night and other quiet periods calves show typical cycles of sleep and wakefulness lasting about one hour, e.g. 20 min awake, 10 min drowsy, 25 min deep sleep, 5 min dreaming sleep; then they wake up again. Over a period of 24 hours a month-old calf may sleep for 6 to 8 hours. Calves usually sleep with their heads tucked into their sides, rarely with their neck extended, unless they are forced to do so by the constraints of a crate or narrow pen. Anyone who has observed a dog 'dreaming' has observed REM sleep. The calf does much the same, in a less dramatic fashion. The muscles of the neck and limbs twitch and the eyeballs can be seen to be in rapid movement under the closed lids. It is an essential requisite of a satisfactory husbandry system that it allows calves to sleep in comfort.

Social behaviour

CURIOSITY

Cattle are timid creatures but are possessed of an insatiable curiosity.

Faced by a new experience like the presence of a novel object in the pen, a calf will approach it carefully, showing clear signs of curiosity (such as sniffing) but braced for flight at any moment (Fig. 22). If the object appears to present no threat, the calf will usually become bolder and actually touch with its nose and tongue the object which may be a bucket that wasn't there yesterday, or a quiet and friendly stockman or even the farm cat. How far the investigation proceeds thereafter depends on how much the object can tolerate. I shall discuss later the extent to which one can use this behaviour pattern (curiosity overcoming fear) to assess the effect of different rearing systems on fear and stress in calves.

PLAY

All young mammals engage in play. Play activities are usually harmless versions of serious adult interactions like fighting and sex. Serious-minded behavioural scientists would have us believe that young animals at play are rehearsing forms of behaviour essential to the adult. It may just be that they are having fun. Young calves at play go through most of the rituals of fighting such as pawing the ground, butting and pushing with their heads. They also run races, although to nothing like the same extent as young lambs. Both male and female calves indulge in precocious sexual behaviour and mount each other quite frequently. The frequency of mounting is increased if calves are implanted with anabolic sex steroids like testosterone or oestradiol, which suggests that mounting behaviour in these young animals is specifically associated with feelings of sex.

Calves under about three months of age do not appear to form stable social hierarchies (or whatever would be the bovine equivalent of a pecking order). At this age it is very unusual to see a dominant calf systematically bullying others by keeping them away from a nipple or food trough. I have also never seen a young calf seriously butting or pushing another calf that is too weak to fight back, presumably because there is no satisfaction in it. Bullying does, however, develop with age and can sometimes be a serious problem with older beef or dairy cattle in yards.

Rearing systems and the development of behaviour

This section is based on two studies, a preliminary investigation conducted by Claire Saville while at the University of Bristol and a similar, larger trial conducted by officers of the U.K. State Veterinary

Fig. 22. Exploration of a novel object. (a) This calf is confidently approaching a novel object in familiar surroundings, (b) This calf is both curious and apprehensive. Note the position of the legs.

Service on 70 commercial calf-rearing units. They were undertaken to investigate the extent to which the rearing system imposed on the animals affected the development of their behaviour from birth to 14 weeks of age and so provide some hard evidence upon which to base any discussion as to the extent to which a particular system of husbandry may meet, or fail to meet, the reasonable behavioural needs of the animals.

The rearing systems that were investigated were:

Group A. *Suckler beef cows with calves at pasture.* This group was taken as the nearest possible approximation to the 'normal' environment for a young calf.

Group B. *Early weaned calves, individual pens.* Calves taken from their mother after receiving colostrum and raised in individual pens with twice-daily bucket-feeding to weaning at 5–6 weeks of age and thereafter in groups in follow-on pens.

Group C. *Early weaned calves, group-reared.* Calves taken from their mother after receiving colostrum and raised in groups with free access to a nipple feeder dispensing either warm, sweet, or cold, acidified milk replacer. Weaning was at about six weeks.

Group D. *Veal calves, individual crates.* Calves reared for veal in individual crates 65–75 cm wide, fed milk replacer twice daily and given no access to bedding or solid food.

Group E. *Veal calves, group-reared.* Calves reared in groups for veal in straw yards. The animals got free access to milk replacer from a nipple dispenser but no solid food other than that which they picked up from the bedding.

In some of the preliminary trials, systematic observations of behaviour were made using teams of observers over continuous periods of 24 hours. In the major trial conducted by the State Veterinary Service, observations were made over four periods each of one hour:

(1) Early morning, to include the morning feed (where applicable)
(2) Late morning, calves relatively undisturbed but normal farm activities
(3) Early afternoon with the whole farm as quiet as possible (ideally the men's lunch hour)
(4) Late afternoon, before (but not including) the evening feed (where applicable)

Observations were made, so far as possible, on the same animals at 2, 6, 10 and 14 weeks of age.

RESTING BEHAVIOUR

The effect of the rearing system on resting behaviour is shown in Table 17 which presents average values for the time spent standing idle (i.e. standing up but doing nothing else), lying on the brisket, lying flat out on the side with one or more legs extended and sleeping. Across all rearing systems the time spent in the most common lying position (on the brisket) declined with age. In general, differences between systems were small and insignificant. The only difference of note concerns veal calves at two weeks of age in individual wooden pens who lay down far less than other calves at that age and, consequently, spent more time standing idle. These calves were on wooden slatted floors with a gap width of 2–3 cm. Moreover, the slats were usually slippery and the calves appeared to be reluctant to shift their position. It is also possible that they may have suffered from cold draughts when lying down on these unbedded slats. In either event, their behaviour at this time was abnormal.

On average, calves that could do so spent less than 2% of their time lying flat out on their flank with their legs extended. In the case of early weaned calves, the period was usually less than 1%. The calves at pasture lay on their sides the most. This is because they were born in the spring and many of the observations were made on warm days in late spring and early summer when the calves would lie stretched out in the sun. Veal calves in yards tended to lie out flat for longer as they grew older. Even though most of the observations on veal calves were made during the winter, the association between time lying flat and air temperature confirmed that the veal calves adopted this position to keep cool. The veal calf in a crate 65–75 cm wide cannot, of course, lie on its side. Compared with a conventionally reared, early weaned calf which adopts this posture less than 1% of the time, this would seem to be a minor inconvenience. However, it is fair to say that the veal calf in a crate cannot adopt the posture it would choose to dissipate heat in circumstances where it feels too warm. Crate design in a conventional veal house therefore contributes to the thermal discomfort that these animals can experience.

The amount of time that the calves appeared to be sleeping also declined with age. Veal calves in yards spent the most time asleep (during the day), about 17% on average. Veal calves in crates spent

Table 17 Resting behaviour of calves in different rearing systems: average values for the percentage time spent standing or lying based on 4–hour observations

	Age (weeks)	Rearing system					Average for all systems
		Suckler calves	Early weaned		Veal		
			Individual	Groups	Individual	Groups	
Lie on brisket	2	46	55	56	40	52	52
	6	45	48	52	46	51	48
	10–14	39	38	45	47	50	42
Lie flat	2	3.8	0.2	1.3	nil	1.0	1.6*
	6	4.9	0.5	0.8	nil	2.0	2.0
	10–14	2.4	0.5	0.8	nil	2.7	1.6
'Sleep'	2	17	17	16	13	22	17
	6	17	9	10	8	13	11
	10–14	9	6	9	7	16	8
Stand idle	2	14	20	18	34	15	19
	6	9	17	14	20	18	15
	10–14	10	12	12	19	16	12

*Excluding individually crated veal calves

the least time asleep at 2 weeks of age. At 14 weeks of age they spent the same time asleep as early weaned calves or calves at pasture. However, by about 10 weeks of age, veal calves in crates were unable to adopt the most common sleeping posture of normal calves, i.e. with their heads tucked into their sides. This may explain why they spent less than half as much time asleep as veal calves in yards. As I wrote earlier, the right to adopt a normal, comfortable sleeping position seems to me to be an unarguable case. The conventional veal crate denies that right to calves in the last four weeks of their life.

ORAL BEHAVIOUR

Table 18 summarises the oral behaviour of calves in the five rearing systems. Calves at pasture spent the longest time eating, mainly because it takes longer to consume the same amount of dry matter when it is grazed in the form of wet grass than when dry, or when concentrated feed is presented in a manger or trough. In this series of observations, calves bucket-fed in individual pens up to weaning and then turned into follow-on yards spent more time eating hay or straw and less time eating concentrates than those raised on teats prior to weaning. I don't want to make too much of this. There may have been differences between these commercial units in amounts and times of concentrate feeding that we have not recorded.

The observations of rumination suggest that the development of rumination behaviour in early weaned calves is closely similar to that of suckled calves at pasture, being observed for about 15% of the time by about 10 weeks of age. Calves bought-in from markets ruminate for long periods from the day of arrival, which suggests that they have eaten considerable amounts of bedding in transit for want of anything better. Veal calves in straw yards ruminate about 10% of the time at 6 weeks of age but thereafter ruminating time declines. This reflects the fact that they eat quite a lot of straw in the first few weeks of life but thereafter take in few solids if bedding is all that is on offer. The rumination behaviour of veal calves in individual crates is extremely variable. On average these animals ruminated for about 6% of the time at an age of 6 weeks or over. A lot of these animals had been on straw bedding at some time in their lives, either in transit to the rearing unit or for the first few days after arrival before being installed in individual crates. The amount of straw picked up during this time appeared to be

Table 18 Oral behaviour of calves in different rearing systems: average values for percentage time spent in different oral activities based on 4–hour observations

	Age (weeks)	Rearing system				
		Sucklers' calves	Early weaned		Veal	
			Individual	Groups	Individual	Groups
Eating hay, grass or straw	2	9	6	6	nil	7
	6	18	10	9	nil	3
	10–14	25	18	10	nil	4
Eating concentrates	2	nil	4	3	nil	nil
	6	1	9	10	nil	nil
	10–14	1	10	15	nil	nil
Ruminating	2	8	10	10	2	5
	6	13	16	17	5	9
	10–14	15	17	17	7	7
Drinking milk	2	5	3	5	1	3
	6	5	1	3	1	3
	10–14	5	nil	nil	2	3
Grooming	2	6	6	5	15	5
	6	6	7	6	11	10
	10–14	8	10	7	12	15
Chewing, licking pen (etc.)	2	1	4	1	14	3
	6	nil	3	1	14	6
	10–14	nil	3	1	16	4
Average total oral activity		44	48	45	33	28

sufficient to stimulate proper rumination for several weeks afterwards. The few calves that got absolutely no access to solid food only showed pseudorumination, and for less than 2% of the time.

Veal calves, whether in crates or yards, spent more time grooming themselves than other calves, especially at about 10 weeks of age. Veal calves get very hot and sweaty at about this time (p. 80) and shed their birth coats. Skin hygiene is poor and infestations of lice are common. The conventional veal crate severely restricts the area of its skin that the animal can reach with its tongue and so, once again, embarrasses normal behaviour. The amount of time early weaned calves spent grooming themselves is similar to that of calves at pasture, which is a little surprising as the latter can usually rely on regular grooming from their mothers.

The amount of time spent in appparently purposeless oral activity like chewing or licking pens, walls, tree stumps, fence posts or other inedible or unswallowable objects showed some very interesting differences between the different rearing systems. Calves at pasture spent some time (1%) in these activities when very young and then lost interest. For once I think it is safe to explain this in terms of motivation. At first, they were curious and explored the environment with their mouths. Having decided that things like fence posts were uneatable and inherently uninteresting they ignored them thereafter. Early weaned calves reared in groups with access to a nipple feeder behaved very similarly to calves at pasture, presumably because they received comparable oral satisfaction. Veal calves in straw yards and early weaned calves in individual pens spent 3–6% of the time in purposeless oral activity. Veal calves in individual crates spent 14–17% of the time chewing or licking the woodwork of their pens, causing, incidentally, a great deal of destruction to the pens. It is not uncommon for a veal calf to lick and nibble a hole right through a ½-inch thick plank during its 16 week confinement. The question arises, 'Is this purposeless oral behaviour of veal calves in crates evidence of stress, manifested by compulsive behaviour?' (see Fraser, 1980; Kiley-Worthington, 1977). Here the business of assessing motivation becomes much more difficult. One criterion of compulsive behaviour is that animals spend an abnormally long time doing it. Veal calves in crates certainly expend an exceptionally long time in oral behaviour that appears to us to be purposeless, but the *total* amount of time that they spend in oral activity (33%, Table 18) is less than that of more conventionally reared animals. This does not suggest neurosis to me. I think it is fair to conclude that they chew the pens because they are bored rather than stressed.

On the whole I believe that the chronically stressed calf tends to become very passive and we have some evidence to suggest that the calves which do most damage to their pens are the strongest, fastest growing animals. Rapid growth is not, on its own, a sufficient criterion by which to establish absence from stress but it does seem to me that purposeless oral activity like pen chewing is more likely to reflect a successful adaptation by an active and curious animal to a barren and unsatisfactory environment, than otherwise.

SOCIAL BEHAVIOUR

Table 19 lists the average time spent by calves in the different rearing systems in a variety of social activities. The amount of time spent in locomotion (walk and run) (which is sometimes only social behaviour) was similar for calves in all groups. Veal calves in crates do pace back and forth and occasionally the younger veal calves were seen to run forward and kick up their heels in a similar way to calves at pasture. They are of course denied any form of social locomotion like running races. The amount of time spent in play, mock fights and mounting was the same in all groups, again with the exception of the veal calves in crates who were denied this activity too. Calves at pasture groomed each other less than other groups but they were, of course, regularly groomed by their mothers, although this activity was not recorded. The increasing incidence with age of cross-grooming by veal calves in crates simply reflects the fact that this was impossible until the calves were tall enough to lick each other over the tops of the crates. The incidence of cross-sucking was low in all groups and cannot be said to have constituted a serious problem on any of the farms under observation. The cross-sucking in the crated veal calves was restricted to the kissing behaviour illustrated in Fig. 21. In the case of the early weaned calves, there was no evidence to suggest that individual penning or group-rearing before weaning had any effect on the incidence of cross-sucking after weaning.

RESPONSE TO MAN

Table 20 summarises the responses of calves in different rearing systems to the following stimuli applied in systematic fashion by a trained observer:

(1) Quiet approach to the calves

Table 19 Social behaviour of calves in different rearing systems: average values for percentage time spent in different social activities based on 4–hour observations

	Age (weeks)	Rearing system				
		Sucklers' calves	Early weaned		Veal	
			Individual	Groups	Individual	Groups
Walk or run	2	10	3	6	4	5
	6	9	4	6	7	6
	10–14	10	6	6	6	4
Play, fight or mount	2	1.6	0.8	3.6	nil	1.3
	6	1.8	1.0	1.7	nil	1.8
	10–14	1.0	1.2	1.3	nil	1.8
Groom other calf	2	0.2	0.9	0.8	nil	1.0
	6	0.2	0.9	0.6	0.2	1.4
	10–14	0.3	1.2	0.6	1.9	1.0
Suck other calf	2	1.0	1.0	1.1	2.0	0.4
	6	nil	0.7	0.6	0.6	0.7
	10–14	nil	0.4	nil	1.0	0.3

Table 20 Response of calves in different rearing systems to the presence of man

	Quiet approach	Sudden sound (clap)	Sudden movement	Novel object (Balloon)
Overall response of calves (%)				
Attraction	6	17	12	63
Little response	88	72	39	34
Alarm	6	11	49	3
Average reaction, by age*				
2 weeks	0	+ 4	–29	+51
14 weeks	+ 5	+ 9	–44	+73
Average reaction, by system*				
Sucklers' calves	–12	– 2	–62	+39
Early weaned, individual	+ 2	+ 3	–43	+53
groups	0	+25	–12	+80
Veal, individual	+19	0	–38	+73
groups	+14	– 7	–36	+80

*Scored response: +100, all calves show attraction; 0, no response; –100, all calves show alarm.

(2) Sudden sound (hand clap)
(3) Sudden movement (throwing arms in the air)
(4) Novel object (placing a balloon in the calf pen)

The study was done on commercial farms and the observations were made by visitors to the farm, i.e. strangers so far as the calves were concerned. Quiet approach to within about 2 metres of the calves produced in most cases (88%) little response, equal numbers (6%) were attracted and approached the observer or backed away. There was no obvious effect of age on this response but, rather as expected, sucklers' calves at pasture were more timid than other groups. Sudden sounds made little impression on the animals. Sudden movement, in this case achieved by the observer suddenly throwing both arms into the air, did produce some alarm which was again greatest in sucklers' calves. There were no obvious differences between the other groups. Most calves showed a positive interest in the sudden appearance of a novel object in the form of a balloon. The degree of interest shown tended to increase with age. The suckler's calves at pasture again obtained the lowest score for this response, although whether this was because they were more timid or because they had more competing interests one cannot say.

The general impression from Table 20 is that, in this study, all forms of artificial rearing systems were rather similar in terms of the effect they had on the response of calves to man. The effect was, moreover, as expected: calves which are in regular contact with man tend to become tamer. It must be said that the farms which were selected for and elected to co-operate in this study were all farms with a high standard of stockmanship.

There was no suggestion from this study that veal calves in crates were unusually fearful in the presence of man. This conflicts with some of my own earlier observations. When I first started to visit conventional crated veal units, the animals were kept in near darkness and entirely cut off from the sights and sounds of normal farm activity. Indeed, visitors and stockmen were requested not to enter the units unless absolutely necessary so as not to frighten the animals. Sudden movement in such units could severely startle calves in the immediate vicinity, causing them to jump around in their wooden crates and the resulting commotion usually created general panic. In recent years there has been a tendency to open up veal units, mainly to improve ventilation and reduce relative humidity and this has had the secondary effect of exposing the calves to the normal sights and sounds of farm activity. In these circumstances their response to the presence of man was normal. One may conclude that all calves should be given a reasonable exposure to normal farm activities to accustom them to the presence of man and the things he does about the farm. If this can be achieved for veal calves reared in crates, the system itself does not appear to induce an abnormal fear response in the animals. Rearing any animal in darkness and isolation may keep it quiet (although extremely bored) while the darkness and isolation can be maintained; such animals are, however, prone to panic at the slightest provocation.

Further reading

Fraser, A.F. (1980) *Farm Animal Behaviour*. Bailliere, Tindall, London.

Hafez, E.S.E. (1969) *The Behaviour of Farm Animals*. Bailliere, Tindall, London.

Hodgson, J. (1971) 'The development of solid food intake in calves'. *Animal Production* **13**, 15–24, 25–36.

Kiley-Worthington, M. (1977) *Behavioural Problems of Farm Animals*. Oriel Press, London.

Signoret, J.P. (1982) *Welfare and Husbandry of Calves*. Martinus Nijhoff, The Hague.

8 Production systems and economics

The story so far has concentrated on what happens to the calf in the first few months of life, expressed in terms of its physiology, health and behaviour. It is necessary now to consider rearing systems primarily from the point of view of the farmer. This chapter therefore deals mainly with economic aspects of production. It does, however, go into some detail on (i) the practicalities of rearing calves from suckler herds of beef cows (management of suckler cows is another story, Dodsworth, 1972), and (ii) alternative systems of veal calf production, since the majority of the text to date has concentrated on early weaning systems for rearing calves from the dairy herd.

Production of weaned calves from suckler cows

SEASON OF CALVING

Table 21 summarises records prepared by the Meat and Livestock Commission (M.L.C.) of physical performance and financial returns for suckler herds producing weaned calves for sale from upland and lowland farms. These are only average values for those herds recorded by M.L.C. in 1981; the original records also include values for the 'top third' of recorded herds which, in effect, sets targets for the average farmer to aim at.

Spring-calving herds calve ideally in March and April, usually out of doors, and calves are usually sold off grass at about seven months of age when the grass runs out. The system is ecologically sound because the peak requirement for food to support the lactating cow and her calf coincides with peak growth of pasture grass. The inevitable disadvantage of all sensible systems such as this is that it produces a glut of calves at the end of the growing season, so calf price per kg liveweight is low. Autumn-calving herds require more conserved fodder and concentrate feed both for the calves and their mothers (Table 21). On most farms it is also necessary

Table 21 Physical performance and financial returns for upland and lowland suckler herds calving in spring and autumn. (Meat and Livestock Commission, 1981)

	Lowland herds		Upland herds	
	Spring	Autumn	Spring	Autumn
Physical performance				
Live calves born (per 100 cows)	91	90	87	93
Live calves reared (per 100 cows)	81	83	76	90
Age of calf at weaning (days)	217	244	209	303
Weight of calf at weaning (kg)	233	256	231	293
Daily weight gain to weaning (kg)	0.89	0.89	0.92	0.83
Feeding, concentrates to cow (kg)	102	167	92	132
concentrates to calf (kg)	35	122	35	135
Hay, silage and straw (tonnes)	3.3	4.8	3.4	5.7
Financial performance (£/cow)				
Output:				
Calf sales/valuation	217	268	209	304
+ cow subsidy	10	11	48	51
– herd replacements	21	23	26	35
Net output	206	256	231	320
Variable costs:				
Concentrates	16	35	15	31
Hay, silage and straw	53	61	58	60
Veterinary	6	8	6	6
Miscellaneous	3	4	4	5
Total variable costs	78	108	83	102
Gross profit margin (£ per cow)	128	148	148	218

to provide housing or at least suitable feeding and shelter facilities over winter. The producer of autum-born sucklers' calves can, however, wean and sell his calves in early summer when the price is high, or keep them on at low cost until about one year of age and sell a much larger animal. Table 21 shows that gross profit margins per cow are higher for autumn-calving than for spring-calving herds, mainly because autumn-born calves made about 14p more per kg, which more than offset increased feed costs. Whether autumn-born calves generate a greater *overall* profit than spring-born calves depends on the magnitude of fixed costs for buildings and labour, which will always be greater for the autumn-calving herd. There is also a suggestion in Table 21 that calf losses tend to be lower in the autumn, which is not surprising since in early autumn, beef cows at grass are usually in peak condition and dispersed on dry, clean pastures so the risk of calves succumbing to infectious disease is minimal.

Management of the spring-born calf

The spring-born calving suckler herd is essentially a low cost, low output enterprise and specialist buildings designed to house the suckler cow cannot be justified on economic grounds alone. The traditional, successful upland beef farm has winter pastures for the cows which are well drained to prevent 'poaching', and provide good shelter in the form of trees and the natural lie of the land. So long as they are adequately fed, and sheltered from the most severe winds and driving rain or snow and the ground is reasonably dry under foot, beef cows tolerate winter conditions in the U.K. or even in the most severely cold ranching areas of the U.S.A. and Canada, with little or no distress or loss of efficiency (Webster, 1974). On lowland farms in the U.K., climatic conditions are much less severe than on the hills but stocking rates are higher and winter pastures become very wet so the problem of poaching becomes much more severe and the fields can become a sea of mud, particularly around the feeding troughs. This is not only most uncomfortable for the cows at the time, but it ruins the pasture for the following summer. Lowland beef herds are therefore housed mainly for the convenience of the stockman and to save the grass.

When beef cows are housed over winter there is inevitably a build-up of infection within the building. This presents no particular problem to the cows themselves but the new-born calves are at risk of contracting a wide range of infectious diseases, the most common of which is *E. coli* septicaemia. It is quite common practice for well-meaning farmers to allow their cows to calve indoors where they can (perhaps) be more comfortable and so that they are close at hand in the event that they need assistance in calving, and then turn them out onto clean pastures when the calf is about 24 hours old. Unfortunately this is not good enough to guarantee freedom from *E. coli* septicaemia which can infect the calf within the first few minutes of its life. From the point of view of neonatal infection it is best if beef cows can calve down out of doors on a sheltered, well-drained pasture that has not held cattle all winter and so is reasonably free of parasites. If there is a high risk of obstetric problems because the cows have been mated to a very large bull (Table 2), or if the weather and conditions under foot are very bad indeed, there may be a case for calving indoors. In these circumstances the cows due to calve should be brought into cleaned, disinfected calving boxes and bedded down on fresh straw. This may sound like a counsel of perfection. However, it is too often the case when suckler herds calve indoors in late winter and early spring that the first few calves are delivered without problems, then *E. coli* septicaemia starts to occur, first in isolated cases, but then getting worse and worse, putting each succeeding calf at greater and greater risk.

The other problem with cows in late winter is that they are likely to be in poor, or at best, ordinary condition. In particular, they are likely to have low circulating levels of the fat-soluble vitamins A, D and E and so little gets through to the calf by way of the colostrum. Their immunoglobulin status may also be low at this time. It is good preventive medicine to administer a multivitamin injection either to the cows before they calve or to the calves immediately after birth. The latter technique is the cheaper. Whether or not it is easier rather depends on where calving takes place.

The second problem likely to face the spring-born suckler's calf is scours (infectious diarrhoea), probably viral in origin, but aggravated by (a) overconsumption of mother's milk, (b) poor immune status, or (c) weather-related stresses out of doors. Many modern beef cows, especially those carrying 50% Friesian blood, produce far more milk in early lactation than is necessary for one calf. It is common therefore to see nutritional diarrhoea in sucklers' calves and when cow and calf are out at pasture there is no way you can restrain the calf from drinking too much. Cold, wet conditions make things worse by increasing gastro-intestinal activity, although cold does not of course actually initiate scours. The best way to control the milk yield of a potentially too milky beef cow is to ensure that she does not get access to very lush, young, spring grass until the calf is at least three to four weeks old and its digestive system has started to mature. Diarrhoea in suckled calves can usually be controlled by careful attention to management but this is not always the case. Nor, too, is the remedy always obvious. I remember attending a Cattleman's Conference in Alberta, Canada some years ago. The morning's session had been devoted to veterinary problems and the vets had enjoyed themselves discussing a range of fascinating but somewhat obscure diseases (Bovine leucosis and toxoplasmosis, as I remember). The Chairman of the conference, who was a rancher from the foothills of Calgary, in summing up, expressed his regret that the vets had not devoted any attention to his most serious problem which was, and always had been, calf scours. One of the vets, a too-recent immigrant from Scotland, leapt to his feet:

'There is no reason, sir, that scours should be a problem. It is simply a matter of over-intensification. Cut down your stocking rate and the problem will disappear.' The rancher drew himself to his full height of about six foot seven. 'I'll remember that.' he said. 'My cows calve down on a sixteen section (sixteen square miles) paddock!' It was, I think, his use of the word 'paddock' that was particularly cutting.

By about four weeks of age the spring-born calf should be safe from the risk of viral or bacterial enteritis and ready to derive the maximum benefit from a combination of milk and young grass. Provided that the cow is milking well these should provide all the nutrients the calf needs to sustain an average growth over the summer of about 0.9 kg (2 lb) per day. Vaccination against lungworms should be carried out when the calves are about 8–10 weeks of age, with a booster 4 weeks later. You should seek advice from your veterinary surgeon as to the advisability of vaccinating your cattle against one or more of the diseases like tetanus, black-quarter or black-leg. These diseases are caused by clostridial organisms which live in the soil. Their distribution is patchy, so advice depends on local knowledge. Preventive medication against parasitic roundworms and liver fluke is again a matter for a local decision based on policies for grassland management.

Most spring-born calves are not weaned from their mothers until the time has come for them to be sold. It is not until they are 5 to 6 months of age that their intake of nutrients from grass exceeds that which they get from their mother's milk. Moreover, up to a weight of about 200 kg they still require for optimal growth a greater protein:energy ratio than that which is produced by rumen fermentation (i.e. a source of UDP, p. 39) and the best way to achieve this is to suck milk protein direct into the abomasum.

Weaning the suckled calf at seven months of age presents some emotional stress to the calf although, provided the calf is removed out of sight and sound, the cows are not particularly perturbed by this stage. What is more likely to cause problems is a sudden change in feeding patterns from grass and milk to (say) grass silage and barley, which can cause digestive upsets and also predispose to pneumonia (Martin and others, 1981). The problems of starting the suckled calf onto food in a covered yard or feedlot are outside the scope of this book but they are undoubtedly greater if suckled calves are weaned from their mothers and shipped out from their farm of origin in one go. The rearer of spring-born calves has no real incentive to wean his calves, say, three weeks before sale because (a) it makes more work, and (b) the calves are likely to go back during this time, so he loses money. It would, however, be of enormous advantage to the buyer if his calves had been conditioned to his finisher or store ration before they left their farm of origin, so much so that he should be prepared to pay a premium for such calves. This is one of many cases where it would be better for the rearer, for the feeder and far better for the animals themselves, if both parties could agree to a contract designed to ensure the good health and uninterrupted growth of the animals, rather than leaving their fate to the vagaries of the market place.

Management of the autumn-born calf

Let us assume that calves are born at pasture in late summer and early autumn. Let us also assume that the cows have had enough grass to eat so that they are in very good condition, but not excessively fat. Let us finally assume that the pasture is not deficient in copper, magnesium, cobalt or one of the other essential minerals. In this case you can practically guarantee that if there are no obstetric problems you will get a healthy bunch of calves. Cows at this time are likely to give a lot of milk and young calves may pass rather loose faeces but the risks of their getting enteritis, dehydration or toxaemia are very low because (i) the challenge from infectious organisms is low, and (ii) the resistance of the calf is very high because the colostrum of the autumn-calving cow is so rich both in immunoglobulins and in fat-soluble vitamins.

The calf born in the autumn can be wintered successfully out of doors. Workers at the North of Scotland College of Agriculture have raised suckled calves out of doors with and without shelter on pasture and on slats on a particularly chilly hillside west of Aberdeen. The performance of calves was unaffected by the degree of shelter available and was the same as that of animals reared indoors. This is consistent with the physiological nature of cold tolerance in calves described in Chapter 4. By the time a well-fed, well-grown calf is about three months old it is as cold tolerant as its mother, since its lower thermal insulation is offset by a higher thermoneutral metabolic heat production. I must add, however, that conditions in the North of Scotland trial were sometimes pretty dismal, not when the temperature was very low but when it had been wet for days on end. It was absolutely proper to impose these conditions during the trial, particularly since there were no perceptible lasting effects on performance and this had been predicted in advance from theoretical calculations. However, I would always recommend that out-wintered autumn-born calves should have access to some shelter from conditions of driving rain or snow and that they should have a reasonably dry bed. It also makes good economic sense to keep the winter feed dry.

When autumn-born calves and their dams are housed over winter, several points should be borne in mind. Since neither cows nor calves need any more than minimal shelter and a dry bed, the building should be as spacious and airy as possible. In areas where winter precipitation is relatively low, the animals probably do best in a partially-roofed strawed yard. In high rainfall areas the accommodation may be roofed over entirely but in these circumstances not less than 2% of the roof area should be open, either in the form of a central open ridge or gaps between the roof sheets. As much as possible of the wall areas should also be of space-boarding, or entirely open. Pneumonia is not such a severe problem in suckled calves as it is for bought-in, artificially reared calves, partly because the density of susceptible animals is lower and partly because they have received ample and appropriate immune protection from their mothers. It does occur, however, and so too do the other problems of poorly ventilated, damp, dismal buildings, namely lice, mites and ringworm.

A creep area should be provided which is accessible only to the calves. Calves go into the creep area, which should be bedded down regularly and generously, to feed, to rest and play and to get out of the way of the cows. It should be possible for the calves to eat both concentrates and forages within the creep. If concentrate feeding is restricted to about 1 kg/day then trough space should be sufficient to allow all calves to eat at once. There is, of course, no need to provide sufficient space to allow all calves to eat hay or silage at the same time but it is a good idea to let them eat forage in the creep, away from competition from the larger and hungrier cows.

Many autumn-calving beef herds are now being accommodated in cubicle units to cut down on straw and labour costs. These can be very successful and do provide reasonable, though less than ideal, standards of comfort and hygiene for the cows. In such units however, it pays to provide a particularly generous creep area because the calves are likely to stay there nearly all the time, returning to their mothers only to feed 6 to 8 times a day.

It is still quite common practice to allow suckled calves only restricted access to their mothers. In the traditional Scottish byre, cow and calf were often tethered separately over winter and the calf allowed to suck from the cow only twice daily. In the south-west of England I have seen units where the calves are housed but the cows kept outside, again except for two feeds per day. I have never been able to understand how this complicated and labour-intensive form of husbandry ever came into being. Calves can, after a period of adaptation, drink nearly as much milk per day in two meals as in eight, but risks of abomasal overload are greater (p. 34). Moreover, regular sucking is likely to stimulate a greater milk production from the cows than twice-daily feeding. I have said already that milk and fresh grass constitute an almost perfectly balanced diet for a young calf in summer. Unrestricted access to milk and high-class silage would, in fact, provide sufficient energy and protein for satisfactory growth in autumn-born calves but it is usual to provide calves with about 1 kg/day of a concentrate feed over winter, partly to ensure a sufficient supply of vitamins and minerals, especially magnesium, in which milk is deficient. If calves are restricted to two milk feeds per day and get 1 kg/day or less of concentrate feed their growth will be quite severely restricted.

Spring turnout occurs of course, when the grass is ready. Calves should receive their first vaccination against husk (parasitic lung-worms) 6 weeks before turnout, with the booster 4 weeks later. The ideal weaning weight for calves from the medium-sized breeds is about 220–250 kg since up to this time it has needed the extra amino acids, fats and other nutrients in milk which pass straight into the abomasum. At greater body weight than this, the processes of rumen fermentation can provide the right amounts and right balance of nutrients from young grass. When cattle become big enough to derive sufficient nutrients from grass alone it is more efficient to put the grass direct into the calf than to allow the cow to eat it first before converting perhaps 10% of its gross energy intake into milk. In practical terms it usually pays to keep the calves with their mothers during the first flush of spring grass to get calf weights up to about 250 kg and prevent the cows from getting too fat, then wean the calves in June/July. Thereafter, the calves should be given the best of the grass, both in terms of nutrient quality and freedom from parasites, and the cows used to tidy up behind.

Multiple suckling

Multiple suckling means rearing more than one calf per cow per year and, at first sight, this seems to make sense since many beef cows clearly have the physiological ability to produce more than enough milk for one calf. In practice however, multiple suckling has never really caught on, for several good reasons.

In terms of the major nutrient, energy, it is more efficient to feed the calf direct than to put the food into the cow first (see above). This becomes particularly relevant if, in order to sustain sufficient milk for two calves, the cow's intake of home-grown grass and forage has to be supplemented by purchased foods. The small, hardy Hereford, Aberdeen Angus and Hereford x Angus crossbred beef cows that roam the great plains and the foothills of North America subsist to a far greater degree than many British beef cows on food which they find for themselves; in winter, for example, gleaning the cornfields after the combine harvesters or rustling for dormant grass or standing hay through the snow. Such cows only produce enough milk for one calf but they don't cost a lot to keep. In Britain, the traditional hardy beef breeds like the Galloway and half-bred cows like the Blue-Gray fit much the same ecological niche. On lowland farms, however, more and more beef cows are coming from the National dairy herd. The Hereford x Friesian bull calf is a very valuable beef animal. Dairy farmers therefore mate a large proportion of their Friesian cows, and especially their heifers, to a Hereford bull hoping for a bull calf but inevitably getting 50% heifer calves which are less valuable for beef because they fatten at lighter weights. A large number of these heifers end up as beef cows which have big appetites and a high potential for milk production. Another contributor to the increase in the number of very milky beef cows has been the incentive schemes to encourage cattle farmers within the European Economic Community to stop producing milk for human consumption. The Hereford x Friesian beef cow eats considerably more food than, say, the Blue-Gray and it can be argued that it should rear two calves per year in order to justify its existence. However, if it is managed efficiently, i.e. it is not overfed and it is mated to a large beef sire like the Charolais, it can be equally profitable as a producer of one high quality calf per year.

One popular form of multiple suckling seen on traditional mixed farms operates as follows. A dairy or 50% dairy cow is mated to a beef bull to calve usually in the autumn. After receiving colostrum the calf is then removed from its mother to a calf box, but given access to her twice daily for feeding. The farmer then buys one, two or three more calves, according to how much milk the cow is producing and then rears them to weaning at about eight to ten weeks. Feeding of hay and starter rations prior to weaning proceeds as described in Chapter 3, the only difference being that the cow provides the milk. Having weaned the first set of calves, the farmer then buys in a new batch and starts again. Depending on how milky the cow is, she may be able to rear 6–10 calves per year in this way. Many farmers allow the cow to continue to suckle her own calf throughout, for reasons determined more by kindness than biological efficiency. In fact, the system could never be described as one that would appeal to an accountant, but it is great fun.

The alternative form of multiple suckling is the free-for-all method that allows purchased calves to run at pasture with beef cows and their own calves, picking up milk from whatever cow they can. The bought-in calves should be strong, healthy and, in my opinion, about four weeks old before being turned out to pasture. If they can be reared up to this time indoors on a good-natured nurse cow as described above, so much the better. In any event they should be accustomed to drink from a teat, not a bucket. Turning younger calves out with the suckler herd carries two serious risks: (i) that the bought-in calves will be unable to poach sufficient milk from the cows and fail to thrive, and (ii) that the bought-in calves will carry infection to the home-bred calves. This latter problem can be very serious. In a closed herd of beef cows, calves tend normally to be very healthy because they are not exposed to infection. This, of course, means also that they may have little protection against disease brought in from outside. It is therefore important to quarantine bought-in calves for at least three weeks before introducing them to the suckler herd. Another drawback to this system is that it is best suited to the spring-calving herd and calf prices are highest in the spring.

Husbandry systems for veal calves

Veal, being derived from the French *veau*, means calf meat. In Britain we distinguish between 'bobby veal' which is the cheap meat of the calf killed in the first week or so of life and useful for little but pies and suchlike, and 'quality veal' which is the expensive product that appears in restaurants as escalopes, Wiener Schnitzel, etc., and is the meat of calves raised entirely or almost entirely on a milk replacer diet usually to a body weight of 160 to 200 kg at an age of 14 to 18 weeks.

There is a tendency to equate veal production with post 1950s intensification of livestock farming but it is, in fact, a very old established and ecologically sound aspect of peasant agriculture which used surplus milk from the house cow to rear a fattened calf for special occasions (such as the return of the prodigal son). Because the calf was killed young, it did not compete with its mother for available forage and cereals. Such a calf would probably have been given some solid food but overwhelmingly the greatest proportion of its nutrients would have come from cow's milk, otherwise it would not have been fat enough for slaughter at a young age. The meat of the biblical fatted calf would also have been very white because cow's milk is deficient in iron (c. 4 mg/kg). Most of these calves would also have suffered from clinical iron deficiency anaemia. It is reasonable to conclude therefore that the assumption that veal meat should be white has emerged as part of our unthinking acceptance of our peasant heritage.

During the 1950s, veal production came under the same pressure for intensification as all other forms of livestock rearing. The increasing size and specialisation of dairy herds generated a large number of calves surplus to requirements on their farm of birth. Increasing public demand for butter and cheese as distinct from liquid milk generated surpluses of skim-milk and whey, and improvements in feed technology made it possible to produce milk replacers for calves in an easy-mix powder form at a competitive price. All these factors led to an enormous increase in the production of quality veal, particularly in Holland, France and Italy. Profits were good and so within a very few years a number of large specialist units appeared, designed to process milk powder through calves in the shortest possible time and at the greatest possible profit. Like so many forms of intensive production, this system evolved largely independently of research into calf physiology and behaviour, the only criteria applied being those of productive efficiency.

The rearing system that has now come to be regarded as conventional operates as follows. Calves are transported from their farms of origin into large rearing units where they are placed in small, individual wooden crates about 150 cm long by 70 cm wide, deprived of all access to bedding and solid food and fed twice daily regulated amounts of liquid diet similar in composition to conventional milk replacers except for iron content. Most milk replacer diets for conventional calf-rearing contain at least 100 mg/kg of iron. Iron is an essential element in haemoglobin, the compound in blood responsible for transport of oxygen and carbon dioxide between the lungs and the tissues of the body (which is why iron deficiency produces anaemia) but it is also a component of myoglobin, the muscle pigment that gives beef its characteristic red colour. If calves are reared on a milk replacer diet containing 100 mg/kg iron, their meat is as red as beef. As I have already pointed out, the belief that veal should be white is a historical accident consequent upon the fact that cows' milk is deficient in iron. Since the vast majority of veal is covered either in batter or a sauce, its colour is irrelevant. I am prepared to accept that there might be a small difference in taste between white veal and red veal although I have seen no convincing proof to that effect (colleagues of mine are investigating the matter as I write). The facts of life are that butchers in Europe will not pay premium prices for veal meat unless the colour is acceptable and if they don't, then returns from calf sales are less than costs of production.

This, of course, prompts the question, 'Why raise veal calves at all?' The reasons are the same now as they were in biblical times. Demand for (or the economic returns from) dairy products generates a supply of calves in excess of the demand for dairy herd replacements or for beef, or in excess of the capacity of the land to support beef. This generates only two economically viable options: to kill the calves at birth, or to raise them for a short period of time on skim-milk, whey and fats, thereby recycling other by-products of the livestock industry. If surplus calves from the dairy herd had been killed at birth this might perhaps have been the kinder option, but they were not; they were moved around markets with minimal attention to their husbandry and welfare until they found a buyer (or died). Production of quality veal, provided that it can be done with reasonable humanity, has to be a better option for the calf, for it ensures at least that it will be cared for because of its high economic value (see Chapter 1).

IRON REQUIREMENTS

Bremner and others (1976) have studied the iron requirements of the veal calf. They concluded that a milk replacer diet containing 25 mg/kg of added iron (making about 30 mg/kg in all) was sufficiently high to prevent clinical anaemia unless the calf was severely deficient in iron at birth, but sufficiently *low* to ensure a satisfactory colour to the meat. A wide range of biochemical tests was used to indicate the degree of anaemia but the most sensitive indicator was a drop in appetite. There is thus no advantage to the veal farmer to produce anaemic calves because they drink less, grow more slowly and make less money. There is, however, no reason why the iron content of the milk should be kept constant throughout growth since iron is stored in the body. It is almost certainly better to ensure an iron intake of 40–60 mg/kg powder during the first six critical weeks of life and then drop iron content to 20 mg/kg in a veal finisher diet. The fine tuning of the iron supply to the veal calf is complicated and does not merit full treatment in this book. Suffice to say that it is possible to produce veal carcasses which are acceptable to the British and North American market (although rather pink by French and Italian standards) without inducing signs of clinical anaemia. It is fair to conclude (in the U.K. at least) that deprivation of iron in conventional veal units does not constitute an unacceptable insult to calf welfare. I may add that meat colour in veal is determined almost entirely by the iron content in the diet. Beliefs that the meat could be kept white by restricting movement and keeping calves in the dark have long since been exposed as myths.

As demonstrated in Chapter 7, the conventional practice of rearing calves for veal in individual wooden crates imposes severe constraints on the development of normal behaviour. There is at present no legislation concerning the size of individual calf crates and so, in practice, economic pressures have reduced crate size to the minimum dimensions possible. Conventional veal crates (150 cm long x 70 cm wide) are only just big enough to contain a veal calf in the last weeks of its life. At this time the animal is unable to turn round, stretch its limbs, adopt all normal lying positions or groom itself. Deprivation of solid food denies the calves the opportunity to perform normal eating and rumination behaviour, although their response to this would appear to be more positive than neurotic (i.e. they find something else to do with their mouths).

The defence of this conventional system of rearing veal calves in crates has been the conventional defence of intensive livestock producers, namely that such animals grow faster than any other class of cattle, that individual feeding ensures individual care and early recognition of ill-health and that isolation restricts the spread of infection. There is a measure of truth in all these arguments, although the health of veal calves in conventional crated units has been far from ideal. The susceptibility of the veal calf to disease or sudden death has undoubtedly contributed to the conservatism of veal producers. When veal calves are apparently so susceptible to disease or sudden death, producers are reluctant to try out any changes of husbandry for fear that things might get worse.

Husbandry systems compared

While conventional veal production is nothing like as offensive as some advocates for animal welfare would claim (usually in the absence of any facts which might confuse their prejudices) there are real reasons for believing that in some respects the system goes beyond the reasonable bounds of humanity dictated by man's compassion for the animals in his care. This has promoted considerable interest in the possibility of developing systems of veal calf-rearing which are more appropriate to their health and behavioural needs. One of the most promising of the commercial attempts to find an alternative to conventional veal production in crates has been the so-called 'straw yard' system (Paxman, 1981). This system appeared on superficial examination to combine good new technology with good traditional husbandry since calves are reared in groups in deep straw which serves both as something comfortable to lie on and something to eat. They get their liquid diet free-choice from an automatic dispenser in the natural way by sucking it through teats. Calves are not denied oral activity, freedom of movement or social activity and by these criteria alone the system must be deemed acceptable on welfare grounds. Taken alone, however, these criteria are not sufficient to ensure an overall improvement in veal calf welfare. For any new system to be commercially viable it must either be economically competitive with the conventional crated system or it must be shown to be so much better in terms of health and welfare that it would become possible to eliminate the proven inadequacies of the crated system by legislation.

My colleagues (particularly Claire Saville and David Welchman) and I at the University of Bristol have studied alternative systems of veal husbandry, concentrating to date mostly on a comparison between the conventional crated system and the straw yard, free-choice milk system in terms not only of behaviour but also of economic traits such as growth rate, food conversion, health and production efficiency. The results of the production trials to date are summarised in Table 22. The main differences between the systems were as follows. Veal calves could be taken to greater weights in crates than in straw yards before they achieved the optimum degree of fat cover to obtain quality payments and before the inevitable decline in food conversion efficiency with age made it prohibitively expensive to continue the feeding regime. The difference in slaughter weight between Friesian bulls and Hereford x Friesian heifers was as anticipated. However, Friesian bulls in crates were on average 30 kg heavier at slaughter than Friesian bulls raised in straw yards. Food conversion ratio was also consistently better for calves in crates than in straw yards. Table 22 shows that whatever short-term economic vagaries determine the profits to be derived from veal production, the crated system was, in our opinion, consistently superior to the straw yard system in terms of yield of meat per unit of calf and feed. Unpublished work from Holland has yielded almost identical conclusions. It is necessary to examine in more detail why this should be so.

FOOD INTAKE AND WEIGHT GAIN

Calves in yards drank more per unit of body size up to about 60 days on feed. Thereafter their intake stabilised at a level just slightly higher than the amount drunk by calves in crates offered liquid feed twice a day. The effect of this was that calves in yards tended to grow markedly faster up to about 50 days, which is a major reason why these animals tend to fatten at lighter weights.

Table 22 Performance and health of veal calves in crates and straw yards (Webster and Saville, 1981)

	Crated calves bucket-fed	Straw yard calves teat-fed	
	Friesian bulls	Friesian bulls	Hereford x Friesian heifers
Finish weight (kg)	203	172	161
Days on feed	112	96	98
Killing-out (%)	58	56	57
Milk consumption (kg)	226	210	203
Food conversion ratio (FCR)	1.46	1.69	1.69
Deaths and culls (%)	11	8	8
Treatment courses (%)			
respiratory disease	5	42	22
enteric disease	25	25	12

FOOD CONVERSION EFFICIENCY

The difference in food conversion efficiency between calves in straw yards and those in crates can be attributed largely to the greater food intake of the former. Differences in activity between the two groups are quite small (Tables 17 and 19) and any possible effects of differences in air temperature can be discounted (p. 79). Restriction of the food intake of calves in straw yards to the amount drunk by calves in crates offers the best prospect for improving food conversion efficiency in the system. This may, in future, be achieved by bucket-feeding calves in yards, although this has several disadvantages in terms of health and behaviour (it increases problems associated with 'vices'). The alternative solution, currently under investigation, is to regulate consumption using a computerised milk dispenser.

The other factor mainly responsible for the poorer food conversion efficiency of our veal calves in yards was that more of them suffered from calf pneumonia (Table 22). Nearly all calves which suffered from respiratory disease enjoyed an uneventful recovery with or without treatment. However, the growth of such calves was, on average, set back 10–14 days. This is equivalent to another 30 kg of milk powder or an increase in FCR from 1.46 to 1.65. It is reasonable to conclude that in our experiments and in commercial practice, respiratory disease has been the second most important obstacle to achieving commercially acceptable values for FCR in straw yards.

Enteric disease was, on the whole, about the same in crates and straw yards. In general, it was only a problem in the first two weeks after arrival and no different from that experienced by any rearer of conventional calves.

At the present time I must conclude that the straw yard system, as described by Paxman (1981), is not yet an economically viable alternative to rearing calves in crates although, of course, labour and building costs are relatively low and it is easier to fit into a flexible, mixed farming system. So far as behavioural needs are concerned the straw yard system is clearly superior. From the standpoint of health there is no absolute characteristic of either system that makes it superior. Isolation of calves in crates does not reduce the risk of enteric disease, nor is there any obvious advantage in denying the calves access to solid food. Indeed, this tends to do more harm than good since it increases the risk of abnormal obstruction due to hairballs.

Control of respiratory disease in veal units is particularly difficult because the large amounts of liquid in food, faeces and urine contribute to a high relative humidity. Very few veal units in Britain are designed to ensure good air hygiene according to the principles described in Chapters 4 and 5. Some units in Holland do allow about 15 m^3 per calf and keep relative humidity below 80% but these are extremely expensive and not justified by the small profits possible in the British veal trade. The economics of veal calf production fluctuate rather violently but the following comparison of our Friesian calves in yards and crates, based on Table 22, is a fair reflection of the average position for the last three years.

	Crated calves bucket-fed (£)	Straw yard calves teat-fed (£)
Calf price	60	60
Cost of milk powder	136	126
Cost of straw	nil	3
Income for calf	235	193
Gross profit margin per calf	+39	+4

These figures are not likely to promote a major change to the straw yard system while the crated system remains permissible in law. I hope, however, that further research will reveal alternative systems that are at once more humane, healthier and economically superior to the crated system. Three possibilities appear viable in the light of present knowledge.

(1) Specialist units rearing bought-in calves in individual straw bedded pens allowing full freedom of movement in naturally ventilated, uninsulated houses at a stocking density of not less than 15 m^3/calf. Milk powder would be bucket-fed as now. The calves would have access to solids from straw bedding at least and some more digestible and palatable solid food may prove beneficial.

(2) Specialist units rearing bought-in calves either in controlled environment or naturally ventilated buildings at a stocking density of not less than 15 m^3/calf. Calves would be reared in groups and allowed to drink powdered milk through teats from an automatic milk dispenser which is computerised to recognise each individual calf, allocate it no more than its ideal daily ration, and notify the stockman of any departures from normal consumption. Systems 1 and 2 should be comparable in terms of gross profit margin as defined above.

(3) Small farm units rearing home-bred calves given milk *ad lib.* from teats. Because of the lower incidence of respiratory and enteric disease in home-bred calves, this system should achieve a food conversion ratio superior to that in the present specialist straw yard units, and building design would be less critical. Food conversaion ratio would be inferior to that of calves in crates because of the lack of control of food intake but the system should still be healthy, humane and profitable.

Contract calf-rearing

Increased specialisation in livestock farming has created a demand for the contract calf-rearer who takes calves either from markets, or preferably direct from their place of birth and rears them either to 12 weeks of age or through their first winter before passing them on to a beef-rearing unit. Calf-rearers may either enter into a contract with their subsequent buyers or play the market. Table 23 illustrates the consequences of playing the market in 1981, an average gross profit margin of £21 per head with the top third approaching £33 per head. This table bears comparison with Table 10 (p. 64) which illustrated the costs of different rearing systems at Liscombe Experimental Husbandry Farm. The 'top third' farms have achieved their economic superiority by (i) buying cheaper calves, (ii) slightly reducing mortality rate, (iii) reducing consumption of milk powder down to the incredibly low figure of 9.5 kg per calf. Milk powder consumption within the average group was 16 kg per calf, which corresponds to conventional twice-daily bucket feeding to about five weeks of age.

Table 23 Physical performance and financial returns for calf-rearing (Meat and Livestock Commission, 1981)

	Average	Top third
Physical performance		
Number of calves in group	26	29
Mortality (%)	3.8	1.8
Rearing period (days)	88	93
Weight at start (kg)	47	45
Weight at end (kg)	113	115
Daily gain (kg)	0.70	0.75
Milk powder (kg)	16.4	9.5
Concentrates (kg)	162	164
Financial performance	(£/head)	
Output		
Calf sales	150.46	146.88
Less calf purchases & mortality	89.16	79.69
Net output	61.30	67.19
Variable costs		
Milk powder	11.11	6.28
Concentrates	24.19	25.14
Hay & straw	1.85	1.20
Veterinary	1.96	1.71
Miscellaneous	1.15	0.03
Total variable costs	40.26	34.36
Gross margin per head	21.04	32.83

The low mortality rate and veterinary costs for the top third group, and the impressive average liveweight gain of 0.75 kg/day all point to the fact that this group were rearing good healthy calves despite cutting milk powder consumption to an absolute minimum. The 'top third' for 1981 must have exercised excellent standards of stockmanship and enjoyed a fair element of good luck. I do not believe that Table 23 should be taken as a general indication that optimal milk powder consumption is less than 10 kg per calf. The farmer who rears calves for only 12 weeks has to take his (relatively small) profit quickly. The farmer who keeps beef calves to slaughter weight or heifers to the point of first calving still spends relatively little during the first 12 weeks; another £5–10 spent on food at this time to give the calf a better start may prove a good investment in the long run.

THE '30-DAY' VEAL CALF

A specialist form of contract calf-rearing is the practice of rearing the smaller Friesian bull calves on milk replacer only in individual crates for about 30 days or until they are large enough to be shipped or flown to continental Europe to be finished for veal. A number of people have both made and lost a great deal of money out of this trade. In essence it is identical to the first five weeks of any conventional crated veal system, which means it carries all the worries of that period and none of the stockman's satisfaction in seeing a good, finished animal. However, while the demand from Europe for veal calves persists, the trade is bound to continue.

The economics of disease

The losses due to disease in a calf unit are as follows:

(1) Deaths and culls; loss of calf value with little or (usually) no salvage value for the carcass
(2) Costs of veterinary services
(3) 'Disease supplement'; increase in feed costs resulting from the fact that calves fail to gain weight or actually lose weight during the time that they are sick, thereby increasing overall food consumption relative to weight gain
(4) Loss of time: increasing fixed costs for labour, interest on loans, amortisation of buildings, etc.

So much is fairly obvious. What is important is to assess the relative significance of these different factors. Aalund (1979) analysed the economic consequences of a fairly severe outbreak of infectious enteritis followed by pneumonia in a specialist unit rearing calves to 12 weeks in Denmark and his conclusions have been adjusted to 1981 U.K. prices in Table 24. Section 1 considers ideal conditions where there is no death or disease. Calf purchase price constitutes 67% of total variable costs. Veterinary and miscellaneous costs (transport, etc.) were £300, 2.5% of total costs. Veterinary costs which include one visit from a veterinary surgeon and preventive medicines for the calves on arrival were about £175 or 1.5% of total cost. With calf selling price at £150 per head, gross profit margin became £3110.

In Aalund's unit, respiratory diseases killed 10 calves before 12 weeks of age, another 11 died of the consequences of enteritis

Table 24 The economics of death and disease in a specialist* unit rearing 100 calves

	Weeks 1–4	Weeks 5–8	Weeks 9–12	12-week total	Gross profit margin (£)
Ideal unit, no death or disease					
Calf purchase @ £80/head	8000	–	–	8000	
Variable costs (£)					
food	940	1060	1590	3590	
veterinary & miscellaneous				300	
Calf sales @ £150/head				15 000	+3110
Infected unit (Aalund, 1979)					
Deaths					
respiratory disease	5	4	1	10	
other	10	1	0	11	
Calves recovering after treatment	7	6	1	14	
Costs (£)					
food	830	880	1330	3040	
disease supplement*	20	18	4	42	
veterinary & miscellaneous				550	
calf sales				11 850	+ 218
'Normal' mortality (5%) and disease					
Deaths, total	4	1	0	5	
Calves recovering after treatment	7	6	0	13	
Costs (£)					
food	920	1020	1510	3450	
disease supplement*	20	18	4	42	
veterinary & miscellaneous				550	
calf sales				14 250	+2208

*Disease supplement is the increase in feed costs assuming that each sick calf failed to gain weight for one week.

and septicaemia and 14 calves recovered after treatment. Disease supplement (extra feed) and veterinary costs were increased in total by only £292 but calf losses reduced sales by £3150, reducing gross profit margin from £3110 down to only £218. Section 3 of Table 22 considers a more typical situation with 5% mortality and 13% of calves recovering after treatment. In this case increase in veterinary costs and disease supplement are the same (£292) and still much less than losses due to mortality (£750). What this Table shows clearly is that the veterinary and extra food costs of keeping a calf alive to 12 weeks of age are trivial compared to the losses, at today's prices, incurred with every calf lost and it provides proof, if proof were needed, that in the matter of calf-rearing, good husbandry and sound economics go hand in hand.

Further reading

Aalund, O. (1979) Management of the newborn calf: an attempt at an economic analysis. In: *Calving Problems and the Early Viability of the Calf.* Ed. B. Hoffman and others. Martinus Nijhoff, The Hague.

Bremner, I. and others (1976) 'Anaemia and veal calf production.' Veterinary Record **92**, p. 203.

Dodsworth, T.L. (1972) *Beef Production.* Pergamon Press, Oxford.

Martin, S.W. and others (1981) 'Factors associated with morbidity and mortality in feedlot calves.' Canadian Journal of Comparative Medicine **45**, p. 103.

Meat and Livestock Commission (1982) 'Beef improvement services.' Data sheets for 1981.

Paxman, P. (1981) Veal production. Quantock loose-housed system. p. 95 in *Alternatives to Intensive Husbandry Systems.* Universities Federation for Animal Welfare, London.

Webster, A.J.F. (1974) 'Heat loss from cattle with particular emphasis on effects of cold.' p. 205 in *Heat Loss from Animals and Man.* ed. J. Monteith & L.E. Mount. Butterworths, London.

Webster, A.J.F. & Saville, C. (1981) Rearing of veal calves. p. 86 in *Alternatives to Intensive Husbandry Systems.* Universities Federation for Animal Welfare, London.

9 Stockmanship and calf welfare

To quote again, and more fully, from the Welfare Code for Cattle as revised by the Farm Animal Welfare Council (1983), 'There are two basic requirements for the welfare of livestock: good stockmanship and a husbandry system which fulfils the health and, *so far as practicable*, the behavioural needs of the animals.' If it were not for the clause in italics (mine) this would be an admirable statement; with it, it becomes just another safe pronouncement made by a committee under pressure from all sides and struggling to achieve some sort of consensus. It does, however, place good stockmanship first in its list of priorities and that is a good thing. No matter how good or bad a particular husbandry system may appear to be in principle (or in a book such as this) the success of the system in terms of the performance, health and welfare of the animals ultimately depends on the quality of the individual stockman. This generalisation assumes perhaps greater importance for the calf taken from its mother soon after birth than for any other farm animal.

Stockmanship

What makes the good stockman or stockwoman? I must finally come round to distinguishing between the two on grounds of sex because it is well documented that when it comes to rearing calves, on average, man for man, women do it better. There have been several epidemiological surveys of factors influencing calf mortality which consistently show that individual variations in stockmanship are quite as important as any of the variations in nutrition and environment which have been discussed in this book. Consistently the best stockmen were shown to be the farmer, his wife or his daughter (but not his son!). Hired hands (most of these surveys are American) were consistently inferior but among this group, women were almost invariably superior to men.

The criteria for good stockmanship are (i) a proper understanding of the animals, (ii) a sense of caring, and (iii) sufficient experience based on proper training. The reason why the farmer, his wife and daughter succeed as stockpeople is that they combine, in varying degrees, experience, a sense of caring and a vested interest in the economic success of the enterprise.

This book was written to help stockmen towards a better understanding of how calves function in sickness and in health. This meets the first criterion for good stockmanship but that alone is not enough. Equally it is not enough for the young stockman to learn his trade only as an apprentice, working alongside the farmer or older stockman. He can never, by definition, hope to be better than his tutor unless he also educates himself to understand the principles that underpin good husbandry. The third thing that the young stockman or stockgirl must do is pick up, through formal training, good habits, especially with regard to hygiene and those manual tasks like dehorning and castration where the skill of the operator is the prime determinant of the humanity of the exercise. Most countries run formal training courses and proficiency tests for stockmen. In the U.K., courses are run by the Agricultural Training Board and local education authorities. Proficiency testing in relevant subjects is carried out in England and Wales by the National Proficiency Test Council and in Scotland by the Scottish Association of Young Farmers' Clubs.

How 'a sense of caring' may be acquired or instilled is rather outside the scope of this book but I do not have much sympathy for the old adage that 'a good stockman is born, not made'. The young person with a liking for farm life and a sympathy with farm animals is not a stockman, merely a promising new pupil to the art and science of stockmanship. Of course, if he is not in sympathy with the animals then it is not in his interest or that of the animals for him to continue.

Markets and transport

In most areas of the world with advanced, intensive dairy industries it is 'normal' practice to remove large numbers of calves from their farm of origin within the first two weeks of life, mix them in markets, transport them again to dealers' premises, mix them again, then transport them to a rearing unit – if they are lucky! If they are unlucky and fail to obtain a good buyer because the price is wrong or because they appear less than fully fit, they may return to market for a second or third time until they finally find a buyer or die. One does not have to be a sentimentalist or a sensationalist to argue that this procedure constitutes the greatest single insult to good welfare that cattle are likely to experience in their lifetime, and I am not forgetting the process of slaughter which, properly conducted, can be carried out with little or no distress to the animal. The worst thing about stresses incurred during the transport of calves is that they are not only unpleasant at the time but can also leave the animal with a legacy of chronic ill health.

In the U.K. the welfare of calves in transit is protected by various Acts of Parliament (The Transit of Calves Order 1963, The Transit of Animals Order 1973 and 1975, the Exported Animals Protection Order 1964 and the Agriculture, Miscellaneous Provisions Act 1968). The welfare code *recommends* that 'a calf should not be removed from the farm of birth for at least 3 days unless for suckling by another newly-calved cow or direct to a place of slaughter as a bobby calf'. It further states that a calf 'should not be disposed of to a dealer or to a market before it is 7 days old unless at foot with its dam'. The Transit Orders give veterinary inspectors of the State Veterinary Service the right to inspect lairages, lorries and the like and to ensure reasonable standards of welfare for calves in transit, including the right to demand the humane slaughter of any calf in severe distress. In the U.K. it is also compulsory, for example, to ensure that calves are fed and watered at intervals of not greater than 12 hours unless the total journey lasts less than 15 hours. It is also compulsory for calves to be rested in lairages and examined by a veterinary surgeon before being exported. All these things are very commendable but they are obviously limited in scope because you cannot ensure good stockmanship through legislation.

It is easy to be critical of procedures for marketing and transporting calves but, if that criticism is to be constructive, it is necessary to define precisely where the problems do, and do not, lie. One can safely dismiss at the outset the criticism that transportation, in itself, constitutes a severe stress. There is good evidence that, provided environmental conditions are satisfactory, calves are quite unaffected by transportation as such. Moreover, it is often easier to marshall and load young calves onto lorries than older, more fractious animals. At this stage of their life they appear to be more bewildered than distressed by all that is going on. The real stresses of transit can be summarised quite simply: (1) Cold stress (or, more rarely, heat stress), (2) Hunger, thirst and dehydration, (3) Exhaustion, and (4) Disease.

It is vitally important for all those concerned with the transit of young calves to be good, experienced stockmen. The calf dealer tends to be a good stockman because his livelihood depends on it but it is equally important that those responsible for calves in lorries, ships and export lairages have not only a sympathy for the animals but an experienced eye for signs of thermal distress, dehydration, exhaustion and disease, and a proper knowledge of how to remedy these things. For example, if a calf arrives at a lairage cold and in a state of exhaustion it is more important to allow it to rest in a deep, comfortable bed of straw than to bucket-feed it (or even worse, drench it) straight away. I shall deal with this in more detail in the next section.

However good the standards of stockmanship and hygiene in lorries and lairages, it is impossible to avoid spreading infection between calves in transit. Movement of calves around the country through markets and dealers' premises has been a major factor contributing to the spread of infections such as *Salmonella typhimurium* and *E.coli* and the increasing incidence of strains resistant to a broad spectrum of antibiotics. Diseases contracted by calves in transit are therefore not only an insult to their own welfare but carry a threat to human public health.

What then can we do about markets? As I said earlier, legislation is no guarantee of good stockmanship and some well-meaning legislation is just plain silly. It is, for example, compulsory (in 1983) for a calf exported from the U.K. to weigh more than 50 kg. The intention is to ensure that calves should not be exported until they are strong enough to travel, but whether or not it weighs 50 kg is a totally inadequate description of its stage of development or its state of fitness. All calves for export have to be examined by a veterinary inspector who has received a long and expensive education in animal health. The 50 kg rule is an insult to the veterinary inspector and, incidentally, to the calves who would be better left alone than manhandled over weighbridges at this stage of the journey.

The insults to welfare associated with cold, hunger, thirst and exhaustion are covered by existing legislation and it would be very improper of me to suggest anything other than the fact that nearly all calves are transported with a proper regard for their welfare. Where problems do occur they can usually be dealt with in the first instance by constructive advice from the inspecting veterinary officer. If this fails he has recourse to the law and should have the courage to use it.

The best way to reduce the transmission of disease through markets would be to pass a law prohibiting the movement of calves off their farm of origin under the age of six weeks. It would, of course, be impossible for anyone to prove in law that a calf in market was, say, five weeks and three days old, but in practice, such a law would prevent the sale of calves from the dairy herd until they had been weaned, dehorned and castrated (where relevant). These things would not apply to veal calves; their position would be comparable to that of the existing 30–day export calf (p. 184).

This proposal is such a radical departure from current practice that I don't see the farming industry rushing to bring it about overnight, despite the very real advantages both in terms of calf welfare and public health. The main drawback to such a scheme is that specialist dairy farmers would have to go back to calf-rearing at least until 6 weeks of age and ideally to 12 weeks, by which time the animals would have come through most of the problems of early development. I shall consider the implications for the farming industry of new legislation in the final section of the book.

An alternative, more modest proposal for reform already advanced by the British Veterinary Association is that it should be illegal to move calves through markets more than once in a period of four weeks. This proposal has the merit that it would be easy to enforce; all it requires is that each calf passing through market be marked by clipping or shaving the hair on the neck or shoulder. A clip mark of this sort is temporary but not easily disguised. Such a law would almost certainly increase the number of calves sold very cheaply as 'bobbies' on their first visit to market and killed shortly thereafter, but on balance I think it would reduce suffering in the population of calves from the dairy herd. It would also discourage farmers and dealers from shipping off to market calves that were weak or sick (because the price would be so poor) and that, in itself, would be no bad thing.

Starting the bought-in calf

Ideally, when a farmer buys in calves from a dealer he should know where they have come from and how long they have been in transit. Since the Tuberculosis Order (1964) requires calves less than 14 days old to be marked in an approved manner (i.e. by an ear tag) before being moved off their farm of origin, the buyer can do a little detective work and examine ear tags to determine the origins of his calves. This does not, however, tell him how long ago they left their home farm and what markets they have been through since. Age and weight are also poor indicators of how old his nominal 8–10 day old calves actually are, although with a little experience he can get a better idea by feeling the developing horn buds.

If the farmer has no guaranteed knowledge of the previous history of the calves that he has bought, he has little option but to assume that on arrival they will be hungry, thirsty and tired, their normal digestive processes will be upset and they will be infected with one or more potential pathogens. The farmer should not accept from the dealer any calves that are obviously sick, for example with severe diarrhoea, navel-ill or septicaemia. That is tough on the individual calf but it is of vital importance to the continued health of the group.

The first thing bought-in calves need on arrival is *rest*. As indicated earlier, if calves have spent the previous day or days in groups in markets or lorries it makes little difference if they are kept together for another night before being put into the rearing pens. The reception area for the calves should be well-bedded with clean straw, free of draughts and if there is some way of providing additional heat when outside temperature is below about 5°C, so much the better. There is no need to provide within this reception area the ventilation standards recommended for calf-rearing units (Chapter 5) since the calves are very small and the straw is clean. It is often possible to set up a cosy reception area for calves in an old farm building that is unsuitable for more permanent accommodation because the ventilation is inadequate or because it cannot be mucked out by mechanical means.

We rest our calves for at least 2–3 hours after arrival before we do anything to them at all. They lie in a group in deep straw and infra-red lamps are provided when the weather is very cold. There is a water bowl in the pen but calves rarely drink much from it. If the calves arrive in the morning, they get a first drink of 1.5 litres of an electrolyte/glucose solution at 40°C (p. 125) in the evening; if they arrive in mid or late afternoon they do not get their first drink until the following morning. Their second drink is the same and the third a 50:50 mix of milk powder, at 125 g/litre concentration, and the electrolyte solution. Provided all goes well, fourth and subsequent feeds are of milk powder alone made up in standard concentration and fed as described on p. 49. If calves are going onto a machine or cold milk dispenser we like to restrict them to 1.5 litres/feed twice daily from a teat for two days longer; this may not be essential but we think it worth while.

The reason for this extreme restriction of nutrient intake during the first two days is based on the assumption that each calf has some indigestion and enteric infection on arrival and should be treated accordingly to prevent it developing into a severe case of enterotoxaemia or septicaemia. A strong calf has ample reserves of energy and protein to withstand this period without food but it does, of course, need both the water and electrolytes.

On the morning after arrival we give all calves a multivitamin injection containing A, D, E and some of the B vitamins. The fat-soluble vitamins (A, D and E) are the most important but most proprietary products contain B vitamins too. They do no harm and, of course, in the pre-ruminant state the calf does not manufacture its own. We also begin three days of oral treatment with antibiotics or sulpha drugs active against most strains of *E.coli* and *Salmonella.* There is some resistance within the Veterinary profession to dispensing drugs wholesale to farmers for the treatment of all bought-in calves because of the fear that this will lead to an increasing incidence of drug-resistant strains of these organisms. This is a real cause for concern, but denying drugs to the farmer who has bought the calves is an attack on the wrong target. As I indicated above, the best way to reduce the spread of antibiotic-resistant strains of enteric bacteria would be to control the movement of calves through markets. I believe therefore that bought-in calves should receive appropriate preventive medicine against *Salmonella* and other organisms, although the exact nature of the drug administered is a matter for the individual veterinary surgeon to decide in the light of his local knowledge.

Routine operations

The castration and dehorning of calves is controlled by the Protection of Animals (Anaesthetic) Acts of 1954 and 1964 which deem the following to be operations performed without due care and humanity unless they are carried out under anaesthetic:

(i) The castration of a bull by means of a device that constricts the flow of blood to the scrotum, unless the device is applied within the first week of life.

(ii) The castration of a bull by any means once it has reached the age of three months.
(iii) The dehorning of adult cattle.
(iv) The disbudding of calves, except by chemical cauterisation within the first week of life.

Under the Veterinary Surgeons Act (1966) only a veterinary surgeon or a veterinary practitioner may castrate a bull which has reached the age of 12 months.

CASTRATION

Item (i) refers to the use of the rubber ring for castrating calves – a procedure similar to that which is used with great success in lambs. I have never used this technique for calves and do not intend to begin. With lambs it appears to cause relatively little distress if applied at one day of age but by the time lambs reach one week of age the operation clearly causes them severe and prolonged pain. Farmers are rightly reluctant to castrate calves under one week of age because of all the other stresses that they are likely to experience at this time. There is also evidence that use of the rubber ring for calves can continue to cause severe pain for as long as three weeks after the operation. (Fenton and others, 1958). Other complications have been local septic inflammation and tetanus. I can find no good reason for the use of the rubber ring to castrate calves and plenty of reasons for dismissing the method out of hand.

Male calves are usually castrated at about seven weeks of age, when an anaesthetic is not necessary. Whether an anaesthetic is advisable depends largely on the skill of the operator. Performed swiftly and cleanly, castration without anaesthetic can stress a calf at this age less than prolonged restraint during administration of local anaesthetic and while it takes effect. The two methods available are surgical removal of the testes (the 'knife' method) and the use of the bloodless castrator to crush the spermatic cord and the blood and nerve supply to the testes. I shall not describe the two procedures in detail because these are skills that must be learnt by direct instruction from a competent demonstrator. It is necessary, however, to consider the pros and cons of the knife versus the bloodless castrator. The advantage of the bloodless castrator is that it almost eliminates the risk of infection because there is no open wound. The advantage of the knife is that the testicles are removed. From a welfare point of view it is necessary to consider not only the pain incurred at the time of castration but that which occurs afterwards. Surgical castration that removes the testicles with a sterile knife after thorough disinfection of the skin appears to cause pain for only a short while after the operation. Bloodless castration may seem like a more humane procedure at the time, but the testes remain in the scrotum as a mass of dying tissue which the body has to deal with. This must not only cause local pain but some degree of general malaise. As a rule I would always recommend the use of the knife for castration provided that the operator was trained and fully competent both in terms of the surgery itself and the necessary antiseptic procedures.

The farmer should, of course, always ask himself the question, 'Is castration necessary?'. The reasons for castration are (i) to prevent indiscriminate breeding, (ii) to render the animals more docile, and (iii) to accelerate fattening. All these reasons were sensible when beef cattle were finished at grass over a period of two years or more. Nowadays when the protein and energy value of the diet can be enhanced by concentrate feeding and animals are finished at a younger age, often in confinement, the justification for castration becomes less obvious. On a high nutrient density diet, bulls grow faster and convert food to lean meat more efficiently than steers (Lawrence, 1980); indeed, the use of anabolic steroids as growth promoters in steers is designed to restore (but not yet to improve upon) the growth-promoting effects of the natural male sex hormones. If bulls are isolated from heifers and finished in yards, the problem of indiscriminate breeding disappears. There are problems, both legal and practical, in managing bulls over 10 months of age and nearing slaughter weight, but these are not insuperable. The question of whether or not to castrate should therefore be an economic decision based on the subsequent husbandry of the animals. There is still some resistance in the meat trade to bull beef based partly on innate conservatism and partly on occasional problems of poor carcass quality which can usually be attributed to stresses in the period immediately prior to slaughter which can be overcome by proper management. There are also prospects that it might be possible to render young bulls infertile by immunological means at least for the time that they are at grass, thereby preventing indiscriminate mating but retaining the growth potential of the intact male. The time may come when castration is the exception rather than the rule.

DEHORNING

Disbudding the calves less than 1 week old by chemical cauterisation with caustic potash has rightly fallen out of favour, partly because it is not always successful and partly because it produces pain that is prolonged long after the operation is complete. This means the only effective method is to remove the horn bud under local anaesthetic with an electric cauteriser when the calf is at least 3 weeks old and the horn tip can be clearly felt. Again, this is a method that should only be attempted after personal instruction from a competent operator.

REMOVAL OF SUPERNUMERARY TEATS

Extra teats on an udder are not harmful but they are, of course, undesirable in a heifer destined to become a dairy (or beef) cow. When the calf is about 4 weeks old the udder should be washed and bathed in an antiseptic solution and the supernumerary teats removed with a pair of sharp sterile scissors. There is seldom much bleeding and it is usually sufficient to apply pressure to the site with cotton wool for a couple of minutes and then treat the area with an antiseptic, astringent spray.

'The reasonable bounds of humanity'

The veterinary surgeon has been described as one 'who wrests the sickle from the grisly hand of death only to thrust it into the bloody fist of the butcher'. The language of this sentence is a bit extreme but no-one can deny that it speaks the truth. Anyone who rears calves or is realistically contemplating rearing them has to come to terms with the fact that he is rearing animals for slaughter. I would add that the good husbandman is also doing his best to preserve and improve the species that he rears by improvements in feeding, breeding and environment. I am no vegetarian and while I respect the right of any individual to be a vegetarian if he or she wishes on whatever grounds he or she chooses, I must point out the logical fallacy in the argument that man has no right to 'exploit' animals. Since the beginning of agriculture, man and farm animals have lived together to exploit the basic resources of earth, sun and water in order to survive. Ruminant animals have been particularly valuable neighbours of ours because they can and do exist to a large extent on food which is not directly available to man (they convert grass into meat and milk). If man were to cease to 'exploit' cattle he would first of all have to kill them all off except for a few living museum pieces in zoos and then he would have to think up an economic way to cut the grass! This is a rather flippant answer but then it is a rather silly question. The fact remains that cattle generate real wealth from the land without destroying it in the process. Moreover, cattle country is beautiful country, be it the great plains and foothills of North America, the mountain pastures of the Alps or the Capability Brown parklands of England's green and pleasant land.

It is, I think, reasonable to assume that man and cattle will continue their relationship as long as both species survive. It is equally reasonable to conclude that, as man determines the nature of this relationship, he has an obligation to ensure reasonable standards of welfare for the animals in his charge, which means a reasonable life according to the Five Freedoms outlined in Chapter 1 and a painless, fearless death. The Very Reverend Dr. Edward Carpenter, Dean of Westminster, recently convened a working party to consider the whole question of *Animals and Ethics* (1980). Practically all of this book is worth quoting but I will restrict myself to extracts from the following paragraphs:

83. The degree of suffering experienced by an animal is dependent on its own physiological and anatomical make-up and is totally unrelated to its beauty, its rarity, its economic or its nuisance value. The welfare of the animal must therefore be considered independently of these things.

46. The basic welfare needs of the calf or piglet are neither quantitatively or qualitatively different from those of a puppy or kitten; the welfare of farm animals is assured only so long as the environment provided by man is not outside its ability to adapt without suffering.

50. Diligent human care for the welfare of such animals as are confined in a few advanced zoological parks can act as a touchstone for the welfare standards of the managed (farm) animal. The concern of those responsible is to provide an environment without distress.

The touchstone suggested in paragraph 50 is of particular interest. It would be foolish to suggest that all farmers should run their farms like zoos and open them regularly to the public, partly for reasons of disease control and partly because they would get no other work done! However, any husbandry system that a farmer was prepared, without qualms, to show and explain to the general public would almost certainly have to be an acceptable system in terms of animal welfare.

In this book I have tried to describe, with considerable reliance upon science and none upon anthropomorphism, the principles that can achieve sound economic performance from calves in a way that is compatible with good health and their behavioural needs. When it comes to performance and health, the demands of economics and welfare are as one. It is only in the context of comfort and behavioural needs that conflicts arise and it is here that it may be necessary to protect both the welfare of the animals and the farmer by recourse to legislation. I have discussed this matter in more general terms elsewhere (Webster, 1982). To cite one example: in the U.K. there are at present no legal constraints on stocking density for any farm animal. If animals in intensive units were permitted the freedoms originally suggested by Brambell (1965) – freedom of movement to be able without difficulty to turn round, groom itself, get up, lie down and stretch its limbs – then, for example, laying hens in battery cages and veal calves in individual crates would require 2–3 times the amount of space they get at present. Such legislation would, of course, completely destroy the conventional, highly capital-intensive systems like battery cages and veal crates. I do not include myself amongst those who applaud such legislation since it would inevitably let in more devils than it cast out. Cages and pens are, on the whole, relatively hygienic and safe arrangements and the producer motivated maximally by profit and minimally by welfare, who was forced by law and economics to get rid of his cages or pens, might be induced to rear his animals in a communal squalor much more injurious to their welfare than at present.

More modest legislation, however, might require veal calves to be accommodated in crates no less than 80 cm wide and laying hens to be provided with not less than 650 cm^2 of floor space per bird. The economic effects of such legislation would be two-fold. First, they would increase costs in these intensive systems by about 20%, i.e. to the point where they would be almost exactly competitive with the best of the semi-intensive systems. Secondly, such legislation would, in the short-term, restrict output. In most of the developed world, where agriculture is at its most intensive, the capacity of the agricultural industry to provide food exceeds the capacity of the local population to eat it, and the *cost* of food production in the developed world exceeds the capacity of much of the third world to buy it. This being so, the consequences of restricting output are quite predictable. In the first instance, the price of goods to the consumer would rise more than the 20% necessary to cover increased production costs because the producers would be back in a seller's market. In short, profits to the producer would be higher than at present. This would encourage expansion at a time when the rules under which farmers operate had been changed slightly so that the best of the alternative semi-intensive systems would be economically competitive with the conventional intensive systems. The incentive to farmers to develop less intensive systems that allowed animals more freedom would undoubtedly be reinforced by the fact that, in a time of high interest rates, these systems tend to be less costly in terms of capital investment.

Once production had re-equilibriated according to the new set of rules, the increase in cost should settle down at about 20% (in real terms) and this increase would undoubtedly be passed on to the consumer. Relative to recent increases in costs of petrol and alcohol, such an increase would be trivial. There has been little, if any, organised consumer resistance to an increase in food costs of this nature in order to achieve a real improvement in animal welfare. While the rules that dictate farm animal welfare remain as they are (or are not), then the individual farmer has no option but to be as intensive as his neighbour in order to survive in the free market. Many British and European farmers have expressed to me their serious concern about the extent to which they have had to go to keep up in this free-for-all. Such farmers would welcome a new set of rules, created by law and fairly enforced throughout the European Economic Community, which enabled them to ensure a reasonable living, to preserve the land that is their heritage and to rear their livestock in a way that they, in their own wisdom, could accept as being 'within the reasonable bounds of humanity'.

Further reading

Brambell, F.W.R. (1965) 'Report of the technical committee to enquire into the welfare of animals kept under intensive livestock husbandry systems'. Cmnd. 2836. H.M. Stationery Office, London.

Carpenter, E. (1980) *Animals and Ethics.* Watkins, London.

Farm Animal Welfare Council (1983) *Codes of Recommendations for the Welfare of Cattle.* H.M.S.O., London.

Fenton, B.K. and others (1958) 'The effects of different castration methods on the growth and well-being of calves'. Veterinary Record, **70:** p. 107.

Lawrence, T.L.J. (1980) *Growth in Animals*. Butterworths, London.

Webster, A.J.F. (1982) 'The economics of animal welfare'. International Journal for the Study of Animal Problems. 3: p. 301.

Index

Abomasum, 10–11, 34, 67
Acidified milk replacers, 54, 57
  mild acids, 57
  strong acids, 57
Acidosis, 140
Adaptation, 94
Air movement, 95
Air temperature, 72–5, 95
Alarm, 94
Amino acids, 22–4
'Animal rights', 14
Antibiotics, sensitivity tests, 124
Antiseptic (definition), 114
Appetite, 42

Bacterial colony forming particles (BCFP), 88–90, 112
Barley, 31
'Barley beef', 34
Barley straw, 33
Behaviour, 14, 144–64
  effect of rearing system, 153–64
Bloat, 63, 138, 147
'Bobby calves', 4, 175
Bronchopneumonia, 122
Bucket-feeding, 47–53
Building heat loss, 106

Calcium, 25–27
Calf diphtheria, 120, 133
Calf hutches, 98–100
Calving, 2–5, 166–69
Casein, 35–6
Castration, 193–95
Cellulose, 38
Cerebrocortical necrosis, 120, 134
Cleaning, 113–15
Clearance of micro-organisms, 87
  from the air, 87–91
  from the respiratory tract, 91–4
Clostridium sp., 35, 143
Coccidiosis, 120, 126
Cold tolerance, 78, 171
Colostrum, 9, 29–30, 35, 48, 170
  fermented, 61
Comfort, 71
Computerised feeders, 55, 70
Contract rearing, 182–84
Controlled environment, 112
Copper deficiency, 25, 28
Corynebacterium sp., 123
Creep feeding, 171
'Curds and whey', 35, 57
Curiosity, 151

Dehorning, 196
Dehydration, 124–26
Diarrhoea, 124–26, 168
Dictol, 142, 169
Dictyocaulus viviparous, 142
Disinfectant (definition), 114
Disinfection, 113–16
Drainage, 104, 109
Drinking behaviour, 146
Dry matter of food, 17
Dutch barn, 101–103
Dyspnoea, 119
Dystokia, 7

Eating behaviour, 147
Economics
  effect of disease, 184–86
  effect of season of calving, 165–67
  veal production, 181
Electrolyte feeding, 125, 128
Energy requirement, 40–43
Energy value of food, 19–20
Enteritis, 120, 124–28
Enterotoxaemia, 124
Escherichia coli, 35, 37, 67–9, 123, 126, 168, 190
Evaporative heat loss, 75–6
Exhaustion, 94

Fan ventilation, 111–13
Farm Animal Welfare Council, 15, 187
Fats in milk, 56
'Five freedoms', 14–15, 144
Floor type (and heat loss), 80
'Follow-on' house, 113
Fumigation, 114
Furazolidone poisoning, 120, 127
Fusiformis necrophorus, 133

Glucose, 38–9
Gruel feeding, 49, 53, 60

Hay, 33
Health, signs of, 119
Heat production, 20, 77
Hemicellulose, 38
Hernia, umbilical, 122
'Hidebound', 121, 125
Husk (or hoose), 142, 169, 172
Hydrochloric acid, 34–5
Hygiene, 51
Hypochlorite, 51, 116
Hypomagnesaemia, 143

Impaction of rumen, 139
Infectious bovine keratoconjunctivitis (IBK), 138
Infectious bovine rhinotracheitis (IBR), 120, 132
Insulation
  external, 74, 79
  of houses, 108, 110
  tissue, 73, 79
Iodine, 124

Joint-ill, 122–23

Lactobacillus sp., 37, 140
Lactose, 36
Lameness, 122
Laminitis, 141
Lead poisoning, 120, 133
Lice, 121, 136, 171
Linseed cake, 60
Lower critical temperature, 77–9
Lungs, infection, 91

Macrophages (in lungs), 92
MAD fibre, 19
Magnesium, 26, 172
Maintenance, 20, 40
Maize, 31
Mange, 121, 137, 171
Markets, 188–91
Meningitis, 120
Metabolic body size, 40–41
Metabolisable energy, 20
Milk, 20
'Milkless' milk powders, 59
Minerals, 24–8, 43–5
Monopitch calf house, 103–105
Mortality, 2
Mucociliary escalator, 92
Multiple suckling, 173–75
Mycoplasma sp., 130

Natural ventilation, 109–111
Navel-ill, 122
New Forest Disease, 121, 138, 143
Non-protein nitrogen, 22, 39
Nutritive value, 17

Oesophageal groove, 10, 40, 53
Once-daily feeding, 51–2
Oral behaviour, 145–51
    effect of rearing system, 157–60
    'purposeless oral behaviour', 149

Pain, 119
Pancreas, 36
Parasitic roundworms, 142
Parietal cells, 34
Parturition, 6–8
Pasteurella haemolytica, 89, 94, 130
Pasture grass, 32
Pepsin, 35
Phenols, 115
Phosphorus, 25–27
Physical comfort, 71, 81
Play, 152
Pneumonia, 13, 120
    enzoötic pneumonia, 130–32
Protein, 22–4, 43
    Crude Protein, 22, 43
    rumen degradable protein (RDP), 39–41
    undegradable dietary protein (UDP), 39–41, 170

Relative humidity, 90–91, 96, 106–109
Response to man, 160–64
Resting behaviour, 151–57
Reticulo-rumen, 10
Ringworm, 116, 121, 125, 171
Rumen, 37
Ruminal acidosis, 63
Ruminant digestion, 37–40
Rumination, 148

Salmonella sp., 37, 65, 128
Salmonellosis, 120, 128–30, 190, 193
Scours, 13, 65–9, 120, 124–27, 146
Selenium, 28, 120, 134
Sensible heat loss, 71
Septicaemia, 120, 122–24, 168
Shock, 121, 125
Silage, 33
Sleep, 151
Social behaviour, 151–60
Sodium bicarbonate, 38, 63, 140
Sodium-treated straw, 34
Soya bean meal, 31
Soya flour (in milk powders), 59
Space requirements, 82, 95
Staphylococcus sp., 123
Starch, 39
Starter ration, 49–50, 61–2
Steam-cleaning, 115
Stocking rate and health, 89
Stockmanship, 186
Streptococcus sp., 123
Stress, 94, 145
Sucking behaviour, 145
Sunlight, 100
Supernumerary teats, 196
Sweating, 75
'Sweet' milk replacer, 50, 56

Teat-feeding, 53–6
Temperature
    rectal, 121
    regulation, 12, 72
Thermal comfort, 71
Thermoneutral zone, 77
Thiamine deficiency, 135
Trace elements, 28
Transmission of infection, 85
Transport, 188–91
Trichophyton verrucosum, 135

Urea, 38

Veal, 28, 175–82
    behavioural deprivation, 178
    economics, 181
    iron requirements, 177
    'straw yard' system, 154, 178–82
    'thirty-day' calves, 184
Ventilation, 86–91, 96, 105–13
Vitamins, 28, 43–5, 120, 134, 168–70, 193
Volatile fatty acids, 37–9

Weaning, 46–7, 55
    sucklers' calves, 170, 173
Welfare, 14
    legislation, 189, 193
Whey, 16

Zoonoses Order (1973), 130